MATERIALS, CIRCUITS AND DEVICES SERIES 28

Heat Management in Integrated Circuits

Other volumes in this series:

Volume 2	**Analogue IC Design: The current-mode approach** C. Toumazou, F.J. Lidgey and D.G. Haigh (Editors)
Volume 3	**Analogue-Digital ASICs: Circuit techniques, design tools and applications** R.S. Soin, F. Maloberti and J. France (Editors)
Volume 4	**Algorithmic and Knowledge-based CAD for VLSI** G.E. Taylor and G. Russell (Editors)
Volume 5	**Switched Currents: An analogue technique for digital technology** C. Toumazou, J.B.C. Hughes and N.C. Battersby (Editors)
Volume 6	**High-frequency Circuit Engineering** F. Nibler *et al.*
Volume 8	**Low-power High-frequency Microelectronics: A unified approach** G. Machado (Editor)
Volume 9	**VLSI Testing: Digital and mixed analogue/digital techniques** S.L. Hurst
Volume 10	**Distributed Feedback Semiconductor Lasers** J.E. Carroll, J.E.A. Whiteaway and R.G.S. Plumb
Volume 11	**Selected Topics in Advanced Solid State and Fibre Optic Sensors** S.M. Vaezi-Nejad (Editor)
Volume 12	**Strained Silicon Heterostructures: Materials and devices** C.K. Maiti, N.B. Chakrabarti and S.K. Ray
Volume 13	**RFIC and MMIC Design and Technology** I.D. Robertson and S. Lucyzyn (Editors)
Volume 14	**Design of High Frequency Integrated Analogue Filters** Y. Sun (Editor)
Volume 15	**Foundations of Digital Signal Processing: Theory, algorithms and hardware design** P. Gaydecki
Volume 16	**Wireless Communications Circuits and Systems** Y. Sun (Editor)
Volume 17	**The Switching Function: Analysis of power electronic circuits** C. Marouchos
Volume 18	**System on Chip: Next generation electronics** B. Al-Hashimi (Editor)
Volume 19	**Test and Diagnosis of Analogue, Mixed-signal and RF Integrated Circuits: The system on chip approach** Y. Sun (Editor)
Volume 20	**Low Power and Low Voltage Circuit Design with the FGMOS Transistor** E. Rodriguez-Villegas
Volume 21	**Technology Computer Aided Design for Si, SiGe and GaAs Integrated Circuits** C.K. Maiti and G.A. Armstrong
Volume 22	**Nanotechnologies** M. Wautelet *et al.*
Volume 23	**Understandable Electric Circuits** M. Wang
Volume 24	**Fundamentals of Electromagnetic Levitation: Engineering sustainability through efficiency** A.J. Sangster

Heat Management in Integrated Circuits

On-chip and system-level monitoring and cooling

Seda Ogrenci-Memik

The Institution of Engineering and Technology

Published by The Institution of Engineering and Technology, London, United Kingdom

The Institution of Engineering and Technology is registered as a Charity in England & Wales (no. 211014) and Scotland (no. SC038698).

First published 2015

The Institution of Engineering and Technology
Michael Faraday House
Six Hills Way, Stevenage
Herts, SG1 2AY, United Kingdom

www.theiet.org

British Library Cataloguing in Publication Data
A catalogue record for this product is available from the British Library

ISBN 978-1-84919-934-6 (hardback)
ISBN 978-1-84919-935-3 (PDF)

Typeset in India by MPS Limited

To Nükhet, Eda, and Burcu

Contents

Acknowledgments

I would like to acknowledge my past and present PhD students and colleagues who collaborated with me over the years on various problems related to thermal monitoring, thermal management, and thermal-aware design. These collaborations have helped me plan and envision material for this book and provided many examples and results to be presented as part of the discussions in this book. I would like to acknowledge Dr Rajarshi Mukherjee, Dr Gokhan Memik, Dr Somsubhra Mondal, Dr Jieyi Long, Dr Min Ni, Dr Brian Leung, Dr Song Liu, Dr Yehea Ismail, Dr Semail Ulgen, Dawei Li, Kaicheng Zhang, and Begum Egilmez.

Chapter 1
Heat in integrated circuits and systems

Thermal and power constraints have become significant design factors for high performance (e.g., desktop computers, servers) and small form factor high-end mobile systems (e.g., smartphones, tablets). At present, power density and the resulting thermal effects are the limiting obstacle to further scaling for production of high-performance integrated circuits (ICs) [1, 2]. Thermal effects due to high power densities are forcing chip designers to apply conservative design margins creating sub-optimal results. This performance gap may translate to billions of dollars of loss for a chip manufacturer. Furthermore, spatially inhomogeneous heat fluxes cannot be addressed effectively by uniform passive cooling solutions. They are often over-designed leading to a significant increase in total cost and also need to be supplemented with other forms of active cooling solutions. For large-scale data centers and supercomputers, the heat generated by every 1 W of computation requires power for cooling in the range of 0.3–0.7 W and the trends are on an upward slope [3–5]. While these costs impact operating expenses and carbon footprints for high-performance computing (HPC) systems more profoundly, it is the heat's impact on system performance and reliability that is a critical issue across the board from mobile devices to supercomputers.

Thermal effects are one of the primary limiting factors influencing achievable performance. All major processor manufacturers correlate the maximum expected performance with the thermal design point (TDP) metric. TDP represents the maximum amount of power that can be sustained by the system during reasonably long execution intervals representative of typical workloads, while only short-lived (in the order of microseconds) crossing over this threshold may be allowed. The past few decades have seen continual growth of the TDP. As the absolute TDP kept rising another phenomenon followed. The distribution of power densities of high-performance chips, particularly, microprocessor chips became highly uneven. At the hot-spots, power densities can be as large as 300+ W/cm^2, while the average power density of the chip is normally an order of magnitude smaller [6, 7]. Hot-spots exceeding the safe temperature thresholds (ranging from 80°C to 105°C in different systems) trigger dynamic thermal management (DTM) engines resulting in a throttling of the system. Also, total power caps dictated by the TDP lead to dynamic provisioning of power budgets to individual cores or chips at run-time, causing slowdown in the system.

In addition to dictating the peak performance, temporal and spatial variations of thermal stress accelerate failure rates in HPC systems [8], requiring more frequent checkpointing and hurting performance [9]. Even when hard failures are avoided, significant variations in performance across different hardware components can be observed due to steep thermal gradients. At any given instant, chips in different nodes within a system may be operating at different power states because of their respective DTM modes, resulting in load imbalance.

1.1 Technology trends driving thermal issues

Scaling is arguably the main factor behind why current systems need to consider thermal impacts as the predominant design bottleneck. Heat dissipation and temperature is directly related to power density in ICs. In latest technology generations, while the scaling factor s still applies to the feature sizes and the resulting capacitive load, the supply voltage no longer scales down, resulting in a linear reduction in power by a factor of $1/s$. When we consider the power density at the same scaling factor, it turns out that it scales linearly by a factor of s. This leads to an ever-increasing heat output generated by chips manufactured at each new process generation. This signifies the increasing prominence of thermal concerns in shaping the power/performance/cooling landscape. High temperatures in chips are undesirable and dangerous. There are several reasons for this. In the following, they will be discussed in more detail.

1.1.1 Design closure concerns

Active devices can exhibit significant fluctuations in performance [10] under thermal cycles. This is due to the strong temperature dependence in two key performance parameters of metal oxide semiconductor field effect transistor (MOSFET) devices. The carrier mobility and threshold voltage of MOSFETs both vary with temperature. These two parameters in turn are critical in determining the current flow in MOSFETs and hence, their switching speed, ultimately dictating the delay of complementary metal oxide semiconductor (CMOS) gates in logic circuits. Interconnect resistivity is also sensitive to temperature and this can lead to significant fluctuations in expected delay [11–14].

Dependence of timing parameters on temperature bring about new challenges in verifying complex designs for design closure and not sacrificing a large guard band while achieving this. The models required to completely verify the performance and functionality of circuits are becoming extremely complex. They are becoming computationally intractable which leads to either unacceptable design cycle time or to gross approximations that limit the effectiveness of design analysis. Design tools that are unaware of thermal effects would force chip designers to apply conservative design margins or make inaccurate estimations creating sub-optimal results in timing and power by 30–40%. This performance gap may translate to billions of dollars of loss for a chip manufacturer. Furthermore, variations directly reflect in the yield of ICs (percentage of chips meeting the expected specification after manufacturing) with more manufactured chips not meeting the performance targets. Yield loss affects both the bottom-line of chip manufacturers and the consumers. Every discarded chip increases the cost of those chips that survive the fabrication process.

Electron mobility has four different contributors, each with their respective relationship to temperature: phonon scattering μ_{ph}, surface roughness scattering μ_{sr}, bulk charge Coulombic scattering μ_{Cb}, and interface charge Coulombic scattering μ_{in}. A widely used MOSFET model established the following temperature dependencies for each component [15]:

$$\frac{1}{\mu_{ph}} \propto T^{3/2} E_{eff}^{1/3}; \quad \frac{1}{\mu_{sr}} \propto T^{-1/\alpha} E_{eff}^{2.1}; \quad \frac{1}{\mu_{Cb}} \propto T^{-1} E_{eff}^{-2}; \quad \frac{1}{\mu_{in}} \propto T E_{eff}^{-1} \tag{1.1}$$

$$\text{where } E_{eff} \approx \frac{(V_{GS} + V_{th})}{6\tau_{ox}} \tag{1.2}$$

V_{GS} is the Gate-to-Source voltage of a MOSFET, V_{th} is the threshold voltage, and τ_{ox} is the oxide thickness. The effective mobility is then expressed as:

$$\mu_{eff} = \mu_0 \left(\frac{1}{\mu_{ph}(T, E_{eff})} + \frac{1}{\mu_{sr}(T, E_{eff})} + \frac{1}{\mu_{Cb}(T, E_{eff})} + \frac{1}{\mu_{in}(T, E_{eff})} \right)^{-1} \tag{1.3}$$

The relationship between different mobility terms and temperature is complex, however, if we focus on devices operating at or above 200 K, the trend is consistently an inverse relationship, that is, electron mobility degrades with increasing temperature. This can be approximated in a compact model [16]:

$$\mu = \mu_{nom} \left(\frac{T_{nom}}{T} \right)^{\alpha\_mob} \tag{1.4}$$

where μ_{nom} is the nominal mobility at the nominal temperature T_{nom} and α_mob is a scaling factor.

Threshold voltage of a MOSFET can be expressed as a combination of the flat band voltage V_{FB} and the Fermi potential ϕ_F [17–19]:

$$V_{th} = V_{FB} + 2\phi_F + \gamma\sqrt{2\phi_F} \tag{1.5}$$

where

$$V_{FB} = \phi_{GS} - \frac{Q_{ss}}{C_{ox}} = \frac{kT}{q} \ln\left(\frac{N_A N_G}{n_i^2} \right) - \frac{Q_{ss}}{C_{ox}} \quad \text{and} \quad \phi_F = \frac{kT}{q} \ln\left(\frac{N_A}{n_i} \right)$$

kT/q is the thermal voltage, N_A and N_G are the substrate and gate doping concentrations, n_i is the silicon's intrinsic carrier concentration, Q_{ss} is the surface charge density, and C_{ox} is the oxide layer's capacitance.

The temperature dependence of the two major potentials can be expressed as:

$$\frac{\partial \phi_{GS}}{\partial T} = \frac{1}{T} \left(\phi_{GS} + \frac{E_{G0}}{2q} + \frac{3kT}{q} \right) \quad \text{and} \quad \frac{\partial \phi_F}{\partial T} = \frac{1}{T} \left(\phi_F - \frac{E_{G0}}{2q} - \frac{3kT}{q} \right)$$

Using empirical parameters from CMOS manufacturing process nodes, it can be shown that the net coefficient of temperature dependence will be a negative quantity, indicating that the threshold voltage of a MOSFET will decrease linearly with rising temperature. Temperature dependence of the threshold voltage can also be approximated with a compact model as follows [16]:

$$V_{th}(T) = V_{th\_nom} + \alpha_{th}(T - T_{nom}) \tag{1.6}$$

The temperature dependence of interconnect resistance is also linear and directly proportional with an empirical material-dependent coefficient α_R:

$$R(T) = R_{nom}[1 + \alpha_R(T - T_{nom})] \tag{1.7}$$

At a nominal temperature T_{nom}, each of the abovementioned performance parameters can be evaluated and a typical case delay estimation of a circuit can be made. However, with the increasing spatial and temporal diversity in thermal response of ICs uniform performance estimation is no longer valid. Particularly, when at the same time instant large thermal gradients create delay variation across seemingly identical components and the eventual cause of a possible timing violation is challenging to predict.

One extreme case is the phenomenon of inversed temperature dependence [20, 21]. As discussed earlier, temperature has an inverse impact on both carrier mobility and threshold voltage. The ultimate gate delay behavior, however, is dictated by the specific operating condition of a logic gate, since that determines the dominant parameter in each case. For instance, for a gate operating with a high threshold voltage level that is closer to the supply voltage, the reduction in the threshold voltage with rising temperature will dominate and improve the drain current flow of the gate. This results in an improvement in gate delay. However, when the gate operates in a regime where the gap between the threshold voltage and supply voltage is larger, then the drain current will be dominated by the carrier mobility, which is inversely related to temperature. Hence, with increasing temperature, delay will degrade in that case.

Estimating digital circuit performance and optimizing it under a wide range of temperature conditions to ensure closure without introducing excessive timing margins is therefore a significant challenge. As a result, thermal induced variations in performance parameters have drawn significant attention in recent years with a number of studies on quantifying and mitigating timing violations in delay paths, clock tree networks, and also within microarchitectural blocks [22–28].

1.1.2 Reliability concerns

Temperature induces several types of wear out in ICs, packaging, and printed circuit boards (PCBs). Primary failure mechanisms linked with wear out and thermal stress are electromigration, negative bias temperature instability (NBTI), stress migration, time-dependent dielectric breakdown (TDDB), and thermal cycling [29].

1.1.2.1 Impact of temperature on electromigration

Electromigration is one of the major wear out mechanisms [30–32] in interconnects. Diffusing metal ions are displaced by conducting electrons under high current density, creating voids. As the thickness of the metal interconnect reduces due to these voids, the resistivity of the interconnect will worsen causing an increase in delay. This can lead to timing violations of an otherwise correctly designed and timed circuit. At the locations, where thinning of the metal occurs, the current density can further increase accelerating the wear out and ultimately resulting in breakdown of the metal and a fatal defect. The diffusivity of ions is exponentially dependent on temperature, which makes temperature a primary enabler of electromigration. The relationship can be expressed in terms of a coefficient D_0, activation energy E_a, Boltzmann constant k, and temperature as follows [32]:

$$D = D_0 e^{-\frac{E_a}{kT}} \tag{1.8}$$

Electromigration will in turn impact the overall reliability of a chip, which can be measured in terms of mean time to failure (MTTF) metric for the interconnect specifically [33]:

$$\frac{1}{MTTF_{EM}} = A_j J^{n_j} e^{-\frac{E_a}{kT}} \tag{1.9}$$

where A_j is a coefficient proportional to the cross-sectional area of the interconnect wire and J is current density adjusted with an empirical constant n. The exponential dependence of the MTTF metric on temperature makes thermal state of ICs a major factor in reliability. Studies point out that temperature-induced reliability problems will tend to limit circuit performance in future technology generations [34–36].

1.1.2.2 Temperature induced instability in transistors

NBTI manifests itself as an overall degradation of a MOSFET's quality, particularly, by increased threshold voltage, decreased drain current, and transconductance. It is not a fatal failure as such, however, it can cause a significant deviation from the expected performance parameters leading to timing violations in the circuit. Operating a MOSFET device at elevated temperatures with a negative bias is the main driver of the phenomenon. Thermal conditions exceeding 100°C are typically categorized as high stress. Hence, it is most notable for P-channel metal oxide semiconductor (PMOS) transistors, since they are biased with a negative Gate-to-Source voltage to switch. Elevated temperatures in the presence of the negative bias are believed to lead to interface traps and accumulation of oxide charge [37]. The shift in threshold voltage due to NBTI effect can be correlated with Gate-to-Source voltage V_{GS}, time t, activation energy E_a (~0.9–1.2 eV), and two fitting parameters m and n as follows [38]:

$$\Delta V_{th} \propto V_{GS}^m t^n e^{-\frac{E_a}{kT}} \tag{1.10}$$

A possible definition of NBTI-induced MTTF could be the amount of time required for the threshold voltage to drop to a certain value. Based on this definition the relationship between MTTF due to NBTI effect and temperature takes the following form:

$$MTTF_{NBTI} \propto M_{NBTI}\left(\frac{1}{V_{GS}}\right)^{\beta} e^{\frac{E_a}{kT}} \tag{1.11}$$

where M_{NBTI} is a technology-related parameter and β is called the voltage acceleration factor (~6–8).

1.1.2.3 Mechanical consequences of thermal stress

Thermal stress also induces strain and deformation in on-chip components such as the dielectric layer as well as off-chip components such as the copper bumps connecting chips with PCBs and the packaging components.

Stress migration refers to the failure mechanism resulting from differences in thermal expansion in different materials. Such differences lead to metal atoms to displace (migrate). While electromigration is a phenomenon driven by charge carriers' interactions, stress migration is purely induced by mechanical stress. This failure hence, affects metal interconnects. Stress migration can be modeled as a function of experienced temperature profile T, the metal's nominal temperature (at the time of deposition) T_{nom}, and material-dependent constant m:

$$MTTF_{SM} \propto |T_{nom} - T|^{-m} e^{\frac{E_a}{kT}} \tag{1.12}$$

Over the course of time, the electric field applied across the gate oxide layer causes wear out in the material. Broadly, the wear out is triggered by existing impurities and defects in the layer, which allow minute amount of leakage current through the layer. As the gate terminal is continuously biased, the tunneling current through the electric field leads to formation of new defects. Over time, the defects grow to a point of complete breakdown of the oxide layer. With technology scaling the gate oxide thickness has also been shrinking, further accelerating this effect. Particularly for thin oxide layers under 6 nm, experimental studies indicate that temperature's influence on the material degradation is profound [39]. The relationship between this defect formation process, also referred to as the TDDB and its associated MTTF can be modeled as follows [38]:

$$MTTF_{TDDB} = M_{TDDB}\frac{A_G}{(V_{GS})^{\alpha-\beta T}} e^{\left(\frac{X}{T}+\frac{Y}{T^2}\right)} \tag{1.13}$$

where M_{TDDB} is an empirical technology constant, A_G is the gate oxide surface area, and α, β, X, and Y are fitting parameters.

Finally, the last failure mechanism with close ties to temperature is the thermal cycling phenomenon. Thermal cycling refers to a chip undergoing significant temperature fluctuations over periods of time. ICs undergo thermal cycles during power up and down and reboot. Furthermore, temporal thermal gradients can become steep due to various power and performance management mechanisms, which are capable of drastically shaping power consumption (e.g., transitioning from high activity to idle/sleep states, engaging Turbo Boost states). Also, drastic changes in workload characteristics can contribute to thermal cycles. During each such thermal cycle, heating up and cooling down suddenly causes stress in certain chip components such as the plastic packaging and the solder bumps at the printer circuit board interface. This stress results in deformation in these structures, ultimately leading to fatal breakdown. The effects of thermal cycles with longer periods are better understood. A model for this failure mechanism has been proposed [29]:

$$MTTF_{TC} \propto \left(\frac{1}{T_{avg} - T_{nom}}\right)^{\gamma} \tag{1.14}$$

where T_{avg} is the average temperature during a thermal cycle, T_{nom} is the ambient temperature, and γ is an empirical material-dependent exponent.

1.1.2.4 Temperature's global impact on reliability

The MTTF definitions for individual failure mechanisms can be studied in isolation or combined to assess the severity of the global thermal impact on reliability. Failure rates of a given chip due to each individual mechanism can be added together to obtain a combined failure rate. Another compact method for quantifying temperature-dependent reliability is to utilize a single exponential relationship. This relation is in fact derived from the Arrhenius equation, which defines the temperature-dependent reaction rate k in a chemical system [40]:

$$k = Ae^{\frac{E_a}{kT}} \tag{1.15}$$

MTTF is inversely correlated with the rate of reaction. This relationship indicates that a 10°C increase in temperature roughly halves the MTTF [39, 40].

Device-specific MTTF in relation to operating temperatures has also been studied in other system components. One such study found that if the ambient temperature increases by 5°C over the design specification, disk drives are 15% more likely to fail [41].

1.1.3 Performance concerns

Temperature has two different kinds of impact on performance. The direct impact on any type of IC operating at high temperatures is that, the delay of active devices and interconnect will most likely degrade due to the relationships between threshold voltage, resistance, etc. discussed in the previous section. Hence, at higher temperatures chips will slow down or designers will be forced to apply more conservative margins to account for such variations at design time, so that, even at average temperatures a significant portion of performance would be left unused. Also, leakage power has a strong dependence on temperature. Hence, with increasing operating temperatures, an increasing fraction of the limited power budgets would be wasted on leakage. Especially in large-scale multicore processor chips that are not carefully designed to be thermally uniform, this can lead to starvation of certain sub-components, when a large fraction of total chip-wide supplied power is drained by those cores that are under high thermal stress.

An indirect, but profound nevertheless, impact of temperature is observed in processors equipped with DTM. DTM mechanisms monitor the junction temperature via on-chip sensors and intervene when a temperature threshold is exceeded; in an attempt to avoid or minimize the reliability concerns described earlier. DTM schemes throttle performance to relieve heat accumulation within ICs when cooling alone cannot mitigate the problem. In commercial products [42–46] the intervention occurs often in the form of scaling down the clock frequency and voltage of the chip, called dynamic voltage and frequency scaling (DVFS). While performance throttling is an effective means of maintaining a thermally safe operating condition for ICs, it comes at the price of intentionally slowing down the system. It must be noted that dynamic intervention mechanisms could be engaged in response to a localized hotspot even if the majority of the rest of the IC operated below that peak temperature point. Several thermal management systems, for instance, apply DVFS chip-wide to all cores globally, even if the thermal emergency is triggered by one particular sensor in a single core. Hence, localized hotspots could penalize the overall system and force it to sacrifice a significant portion of compute cycles. As a result, these interventions protect hardware, prevent fatal failures, and minimize cooling energy, but introduce inevitable performance degradations and variations. Variation in performance is particularly detrimental in HPC systems. Two nodes in a representative HPC system can experience different thermal states due to variations in their processors' leakage and access to cooling. Such imbalance can cause one of the nodes to spend 36.38–56.25% more of its run-time on a throttled-down frequency compared to the other node. Furthermore, this throttling could be due to the activity of a single thread for only a fraction of the total execution time. Ultimately, the impact of throttling in this system can lead to an overall performance degradation between 2.9% and 56.9%, presenting a large, unpredictable spread.

1.1.4 Cost- and user-driven, economic concerns

Thermal effects due to high power densities are forcing chip designers to apply conservative design margins creating sub-optimal results. This performance gap may translate to billions of dollars of loss for a chip manufacturer. Furthermore, spatially inhomogeneous heat fluxes cannot be addressed effectively by uniform passive cooling solutions, for example, heat sinks. They are often over-designed leading to a significant increase in total cost. One metric to define the efficiency of a passive cooling solution (including vertically combined heat spreader, thermal interface material (TIM), and heatsink) is an equivalent thermal resistance from junction to the ambient. This can be measured as the ratio of the temperature difference between the heatsink and the ambient air and the TDP power consumed by the processor package. For a lower end (~700 MHz) processor with peak power rated at around 20 W and peak junction to ambient temperature difference of 50°C, thermal resistance lower than 2°C/W is required. The required thermal resistance can be achieved by optimizing the geometry of the heatsink and using innovative materials and designs. However, this will come at a cost. For instance, the prices of heat sinks advertised for the Intel Pentium 4 family of processors, all using the same thermal grease-based TIM, can span a range of almost 3X. Similar material and production cost trade-offs exist for the choice of the TIM. Two common types of materials used in processor packages are thermal grease and phase change materials (PCMs). Thermal paste made of various compounds containing a mix of zinc oxide, silicon, aluminum, and other metal oxides are also used. While the thermal performance of PCMs is superior to thermal grease, the application of thermal grease requires less sophistication and would add less towards the manufacturing cost.

Furthermore, most high-end computing systems use additional cooling solutions, besides heat sinks, ranging from cooling fans in desktop systems and servers to liquid cooling in datacenters. System managers are extremely motivated to battle cooling-energy cost: the largest line item in their total operating cost. Commercial data centers and supercomputing centers are under immense pressure to reduce their cooling cost and carbon footprint [3–5]. This makes thermal effects a first-order design and cost consideration.

Thermal impacts also affect the end user's satisfaction with the computing device. A longstanding non-computing related benchmark of operation quality has been the acoustic effect. Higher quality thermal solutions built into chip packages and sophisticated DTM schemes help to run cooling fans at lower speeds, which in turn reduces acoustic noise.

Thermal-aware design and management trend has also caught on with the other end of the spectrum; mobile handheld devices. Heat pain is experienced by most humans when they touch an object that is hotter than 45°C [47, 48], which is not uncommon in mobile devices. For example, there have been user complaints regarding the outer temperature of a tablet device while playing graphics intensive games and it is reported that the hottest spot on this device reaches 47°C [49]. In fact, complaints of excessive operating temperatures are common to many smartphones [50, 51]. Hence, thermal effects have a far reaching impact on the production and operation cost of high-performance ICs in all domains of computing.

Mobile device manufacturers are investigating introduction of new materials into their products to improve the user's experience and prevent heat-related discomfort. Coating the inner side of a smartphone's cover with encapsulated PCM is a possible option [52]. Depending on the shape of the coated area (if applied to a wide area of the casing, it should be noted that the contour of the casing for different smartphones and tablets can be irregular) this is a new item to factor into design and manufacturing cost.

1.2 Heat dissipation in chips

Power dissipated within an IC is converted to heat. There are two main contributors to heat dissipation at a given location within an IC. First, the power dissipated by active and passive devices produce heat. Second, there exists thermal coupling between each component and all components in its physical vicinity. These could be the ambient air, other components placed on the same board, rack, etc. The power consumption within an IC is comprised of the dynamic power P_{dyn} and leakage power P_{leak}. The most dominant contributor to dynamic power in a carefully designed circuit is capacitive switching, which can be expressed as a function of capacitive load C_L, the load node's activity factor α, clock frequency f, and the supply voltage V_{dd}:

$$P_{dyn} = \alpha\, C_L V_{dd}^2 f \tag{1.16}$$

Leakage power, on the other hand, is consumed through current leakage paths of active devices via subthreshold leakage, p–n junction reverse-bias leakage, gate-induced drain leakage, and tunneling leakage. Subthreshold leakage is the dominant component and it is a function of active device parameters as well as environmental factors such as temperature.

$$I_{leak} = I_S e^{\left[\frac{\left(V_{GS}-V_{th}-V_{offset}\right)}{nV_T}\right]}\left(1 - e^{\left(-\frac{V_{DS}}{V_T}\right)}\right) \tag{1.17}$$

where I_S represents a technology-dependent current coefficient encompassing geometric technology features, V_{GS} is the gate voltage, V_{th} is the threshold voltage, V_{offset} is a lumped voltage term representing secondary effects involving the linearized body effect and the drain-induced barrier lowering effect [53], and $V_T = kT/q$ is the thermal voltage, which involves a dependency on temperature.

Among the major leakage current components, the p–n junction reverse-bias current increases exponentially with temperature, and the subthreshold leakage current, which dominates modern device off-state leakage is a superlinear function of temperature as well through the influence of the threshold voltage V_{th} in (1.17). This creates a circular relationship between power consumption, heat dissipation, and temperature. Higher temperature leads to more leakage, which further escalates heat dissipation and temperature. In an extreme case, where the system is unable to dispose of accumulated heat at a faster rate and drop the operating temperature, this may lead to thermal runaway [54]. An IC, which cannot converge to a steady-state temperature and instead experiences a continuously climbing operating temperature due to avalanching leakage power is said to be in a state of thermal runaway. This would lead to permanent damages to the IC and its package unless power is shut off.

The combination of dynamic and leakage power results in a total power consumption of

$$P_{tot} = P_{dyn} + P_{leak} \tag{1.18}$$

Heat dissipation within an IC is proportional to power density and the heat conduction mechanisms dictate the rate of dissipation, the transient thermal events, and the eventual steady-state temperatures reached. This mechanism can be described by Fourier's Law as follows:

$$q = -k_t \nabla T = \frac{\nabla T}{r_t} \tag{1.19}$$

where heat flux q is proportional to the temperature gradient and the thermal conductivity k_t of the material. In other words, heat flux is proportional to the ratio of temperature gradient and thermal resistivity r_t. A negative sign is conventionally used to indicate heat flux from the higher temperature point of the gradient towards

Table 1.1 Quantities governing the thermal state of an IC

Thermal metric	Unit
Q: Heat	J
P: Heat flow rate	W
q: Heat flux	W/m^2
k: Thermal conductivity	W/(mK)
R: Thermal resistivity	K/W
C: Thermal capacitance/unit mass	J/(kg K)
ρ: Material density	kg/m^3
T: Temperature	K
ΔT: Temperature gradient	K

the lower. In practice, an active IC will be at a higher temperature than the ambient during its transient thermal response and at the same temperature as the ambient at the steady state. Hence, heat is expected to flow from within the IC towards the ambient. Heat flux is equivalent to power density in an IC.

The rate of change in presence of a heat source with power density P_{den} at a point $r = (x, y, z)$ in space can be described as a Poisson's equation as follows:

$$\nabla q = -k_t \nabla^2 T(r) = P_{den}(r,t) - \rho C \frac{\partial T(r,t)}{\partial t} \tag{1.20}$$

ρ and C are the density and the heat capacity of the material, respectively. The solution to the partial differential equation (PDE) in (1.20) yields the transient temperature at a given point. At the steady state, the rate of heat dissipated to the ambient is equal to the heat generated within the IC. This is a fundamental boundary condition assumption applied to thermally stable systems. Therefore, at the steady state all derivatives with respect to time diminish yielding the following PDE:

$$\nabla^2 T(r) = -\frac{P_{den}(r)}{k_t} \tag{1.21}$$

Table 1.1 summarizes relevant thermal parameters along with their units. For a subset of these quantities we also summarize their typical ranges of values for components in commercial systems in Table 1.2. The representative range of maximum power consumption values are based on TDP values reported in processor datasheets of various leading manufacturers. The chip-wide range for power densities have been derived from approximate die sizes of representative chip families [55, 56]. Power densities at localized hotspots [57] can well exceed these average chip-wide values to reach up to hundreds of W/cm^2.

1.2.1 Examples of thermal responses in high-performance chips

Before we discuss the specific mechanisms for accumulation and conduction of heat within ICs manufactured in various technologies (e.g., 2D vs. 3D), presenting a visual depiction of typical thermal response cases would provide some insight. In the following, we present representative temperature responses of

Table 1.2 Representative value ranges for system components in practical settings

Thermal metric	**Range**
P: Heat flow rate	Server-type chip multiprocessors (CMPs) (45–22 nm) Max per chip: 15–300 W Number of cores (2–61) Mobile (notebook, smartphone) (45–32 nm) Max per chip: 4–35 W Number of cores (2–4)
q: Heat flux	10–50 W/cm^2 (chip-wide) local hotspots can reach hundreds of W/cm^2
k: Thermal conductivity	Silicon: 150 W/(mK) Bulk copper: 400 W/(mK) Aluminum alloy: 120–240 W/(mK) TIM (high grade thermal grease): ~8 W/(mK) TIM (phase change material): 3 W/(mK) (20 W/(mK) with high material failure rate)

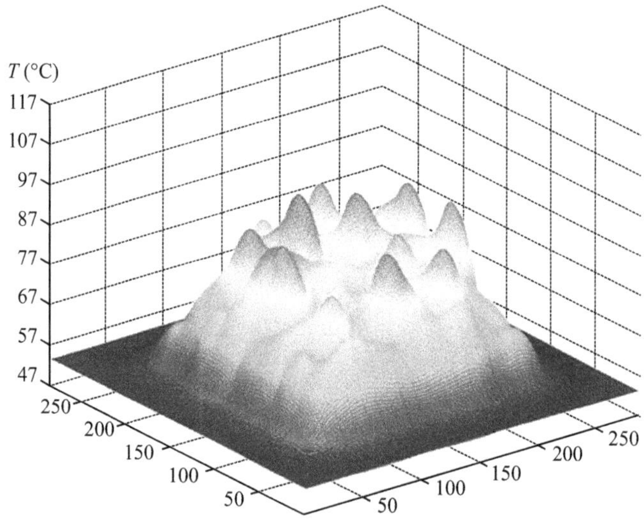

Figure 1.1 Thermal map derived from thermal simulation of a 16-core processor under a heavy workload. Thermal hotspots at diverse locations and significant gradients are present in the system

high-performance chips under different operation scenarios. These illustrate how thermal response is shaped based on variation in activity, physical location of a chip, and presence of dynamic management schemes both in time and space.

Figure 1.1 depicts the thermal map of a 16-core processor derived from thermal simulations. Thermal behavior can vary drastically both temporally and spatially within a processor chip. This image illustrates the temperature distribution within a multicore processor chip under a heavy workload. High-temperature peaks

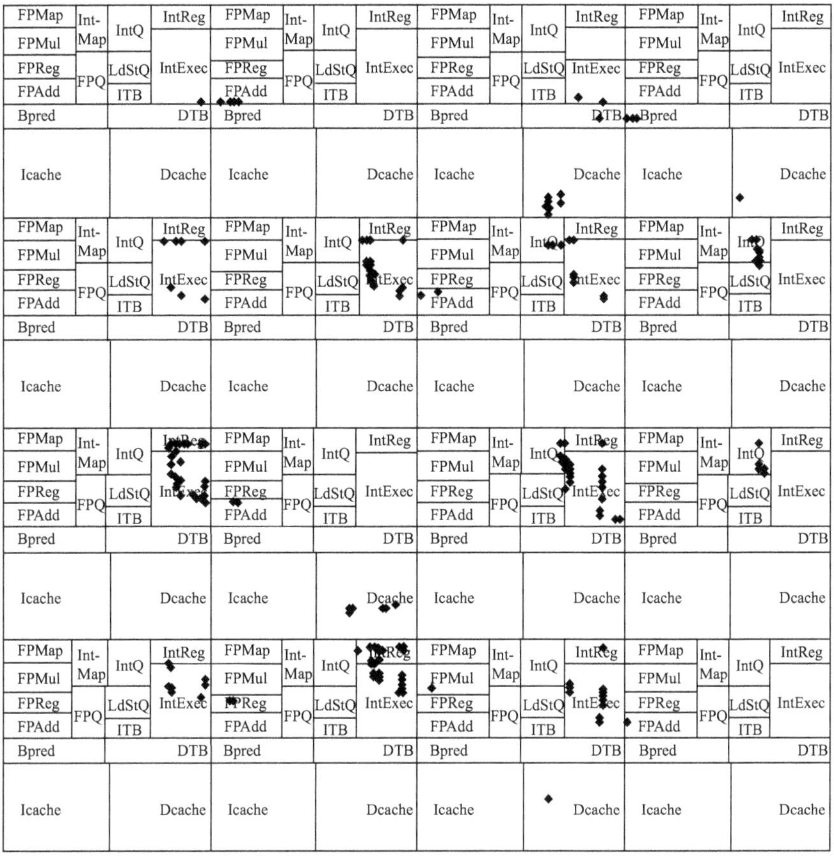

Figure 1.2 Locations of hotspots (temperature peaks exceeding 82°C) observed during the execution of 16 different workload mixes over time on a 16-core processor

denote the thermal hotspots. All modern processor chips are equipped with thermal sensors, which trigger a forced slowdown on the entire chip or parts of it, in an attempt to maintain the temperature of the hottest point on the chip below a predetermined threshold. This inevitably hurts performance, especially when a few localized hotspots dominate the thermal response due to uneven activity distribution across system components.

Figure 1.2 depicts a second example highlighting the variation of hotspot locations among workloads. In this example a hotspot for a given workload is a location where the maximum steady-state temperature exceeds a limit (e.g., 82°C) during execution of that workload. All such hotspots are depicted through 16 workload mixes in the figure. In this case study 667 distinct hotspots from 16 different workload runs with temperatures ranging between 82°C and 117°C were observed. Figure 1.3 shows the same 667 hotspot locations superimposed on the floorplan of a single core. These

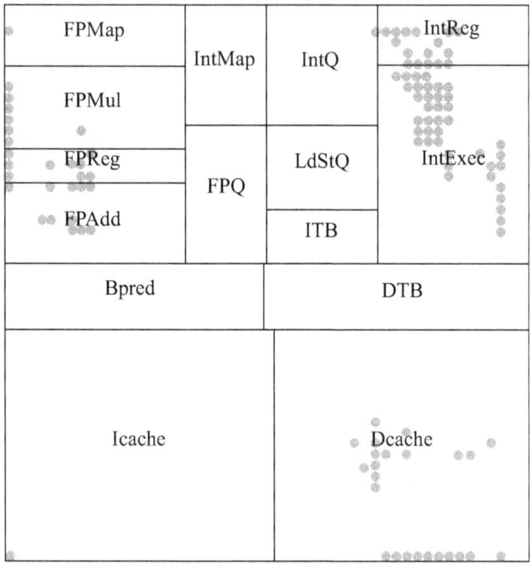

Figure 1.3 Hotspots across different workloads executed on a 16-core processor are folded onto the floorplan of a single core. Thermal response is shaped by two reasons: the workload-dependent characteristics (non-overlapping hotspot locations are observed for different workloads due to this reason) and workload-independent characteristics (the inherent power density of certain units are high regardless of the application)

examples show that (i) hotspots tend to occur in certain sub-blocks of the cores more often than other locations (e.g., integer and floating point execution units, register file, and data cache) and (ii) different workloads create different number of hotspots and these are formed at non-overlapping locations.

The next example illustrates the evolution of a high-performance processor chip's thermal state towards the steady state. The temperature plot in Figure 1.4 is based on data collected from on-chip temperature sensors of an Intel Xeon Phi co-processor chip. The steady-state temperature is not extremely high because this system is utilizing a cooling fan at the fullest speed in this particular example.

Figure 1.5 demonstrates the impact of application activity on the thermal response. This figure depicts the temperature plot of the same Xeon Phi co-processor for two different runs with two different workloads from HPC domain. Both applications reach a steady state on this chip, however, their respective maximum temperatures are different due to the differences in their activities.

The transient behavior of the thermal response is also visible in both Figures 1.4 and 1.5. Figure 1.6 further illustrates the variation of temperature over time during a longer execution time of a periodic task. The warm-up period starts out at a low temperature and during the period of high activity the temperature

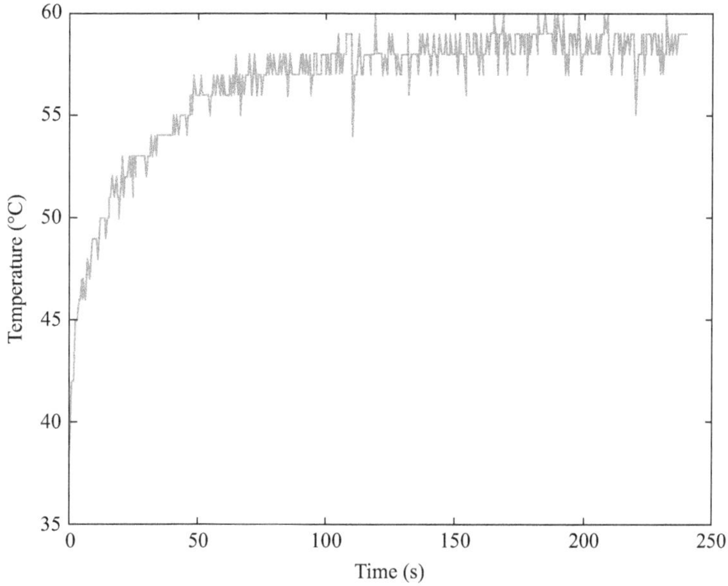

Figure 1.4 Data collected from the on-die digital temperature sensor (DTS) of a Xeon Phi co-processor chip. The temperature of the chip settles into a steady state at around 68°C. This test was executed in a system with a cooling fan operating at the full speed, which ensures that the peak temperatures are well below the critical threshold for triggering performance throttling

climbs. These phases are repeated with corresponding hills and valleys in the temperature plot. Since each phase starts out at the (slightly higher) initial temperature reached at the conclusion of the pervious phase, the consecutive phase tends to climb to a higher peak temperature.

Finally, Figure 1.7 illustrates the impact of DTM. This figure plots both the values sampled from the on-die DTS and the clock frequency information collected by the system. For this particular system the critical temperature threshold to trigger performance has been set by the manufacturer to 102°C. This has been indicated on the plot with a solid horizontal line. Soon after the chip reaching 102°C, the system reports that frequency throttling has started. At first, frequency is dropped for short periods of time. However, the system senses that this does not suffice to drop the temperature back down to 102°C. In response, the clock frequency is kept at the lowest level for a prolonged period of time until the sensors report the temperature dropping to 102°C at around 240 s. Once the sensors report that the temperature is below the critical limit, the aggressive frequency throttling concludes and the clock frequency is increased.

In conclusion, the thermal response of a high-performance chip can be diverse under different settings and be affected by internal (extent of performance throttling, power consumption, layout of the chip, the quality of materials used for its

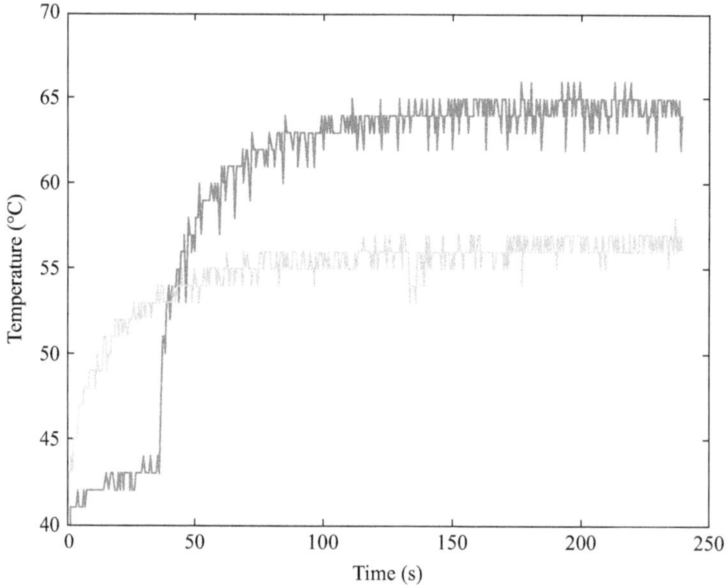

Figure 1.5 *Workload characteristics influence the thermal response. In this figure, the difference between the steady-state temperatures reached by two applications is clearly visible. All system properties (processor, cooling fan speed) are maintained equally between the two runs. The difference is mainly attributed to the variation in compute intensity of the workloads*

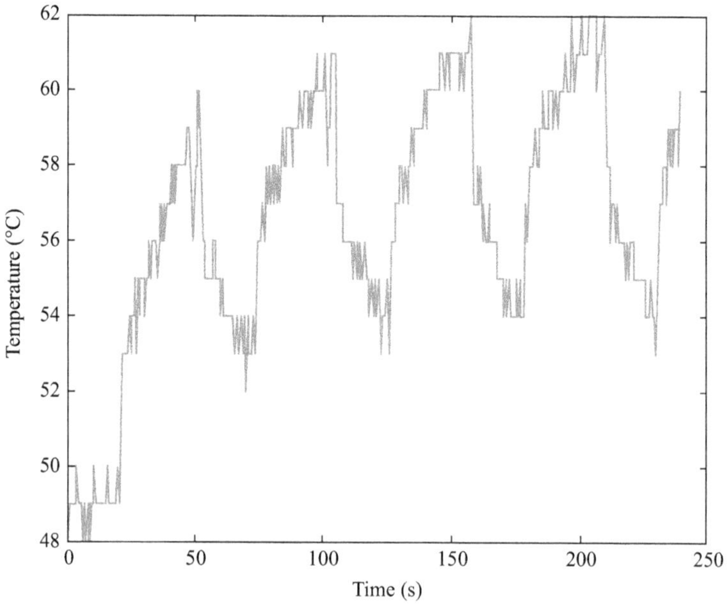

Figure 1.6 *A periodic task executing through phases of high and low activity. The temperature of the processor follows the trend*

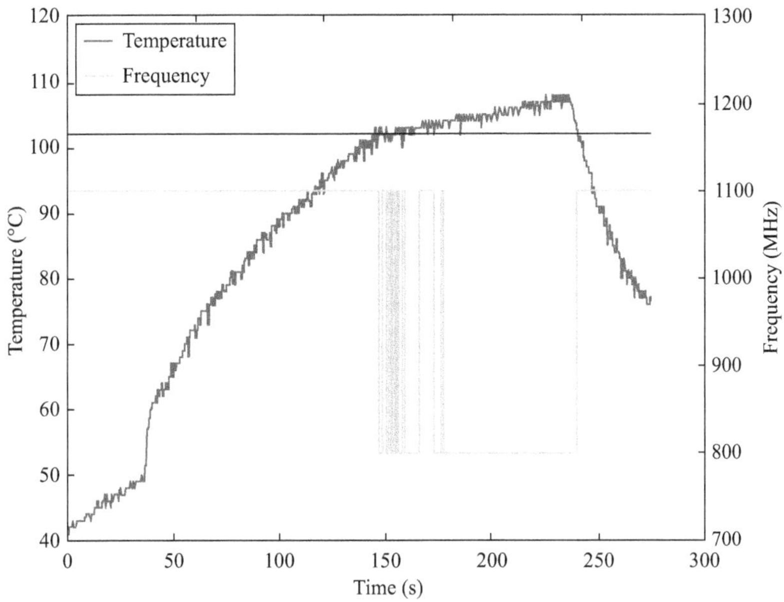

Figure 1.7 Thermal response of the processor under heavy workload and limited cooling: when the temperature reaches the critical limit of 102°C (denoted by a solid horizontal line) frequency throttling starts. The system eventually resorts to dropping to the lowest frequency to bring the die temperature back to 102°C. Upon descending below the safe temperature, frequency is ramped up again

encasing, etc.) and external factors (workload, strength of the passive cooling, active cooling, etc.). The task of on-chip temperature monitors is to track these events and facilitate coordination among automatic interventions of the hardware and strategic decisions at larger time scales by the software.

1.2.2 Heat dissipation paths

Heat dissipation in high-performance ICs and systems are governed by the thermal capacitances and conductivities of individual components and their respective power densities. In the following, we will review heat dissipation paths within IC packages of different technologies (planar, 2.5D, 3D), as well as at the board level.

There are several thermal pathways within an IC through which generated heat is dissipated. Figure 1.8 illustrates a representative configuration for an IC including its surrounding materials and layers. For all practical purposes heat generated from within an IC placed in the die is considered to be the main source. To a second degree other chip packages and modules (e.g., regulators) sharing the same PCB may contribute to each other's effective heat output through thermal coupling. However, the heat map of an IC's interior can be largely defined within the confines of the elements shown in Figure 1.8. There can be variations to the configuration depicted in

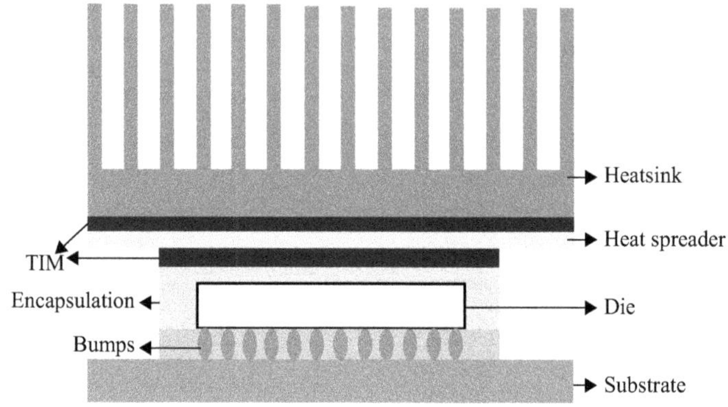

Figure 1.8 Illustration of an IC and relevant layers surrounding the die

Figure 1.8. For example, the die in this figure is depicted as using the flip chip interconnect method (popular choice in processor chips) attached to the package substrate with copper bumps. Some ICs are wire bonded instead. The substrate is then attached with solder bumps or through socket pins to the PCB. Some IC packages may not include a heat spreader and the die would be directly attached to the heatsink with a TIM layer. The configuration without an integrated heat spreader is usually preferred for processor chips in the mobile computing domain, whereas processor chips deployed in server applications and HPC domain often utilize the heatspreader as well. The structure of the heatsink can also vary. The heatsink depicted here contains fins while some may not. Also, heatsinks can contain additional heat pipes to circulate cooling fan air or host additional active cooling structures such as liquid cooling pipes.

The geometry and placement constraints within the casing generally do not allow air circulation over the underside of the PCBs. Furthermore, most inorganic material with high thermal conductance is located on top of the die towards the heatsink, while the layers under the die have drastically limited heat conductance. As a result, along the vertical dimension heat flow is primarily in the upward directions towards the heatsink. Heat conductance also occurs laterally within the die radiating from the locations of the heat sources. However, the heat conductance of silicon along the lateral axis is poorer than the conductance in the vertical axis, which makes the upward path towards the heatsink, the primary path. Nevertheless, detailed modeling and thermal engineering for cooling solutions take the lateral effects into account for thermal coupling between on-chip components.

The most common method to represent the thermal interactions is to utilize the analogy between Ohm's Law and Fourier's Law, which establish the equivalence between transient heat flow networks and electric circuits [58]. Electrical current is the ratio of a voltage differential and electrical resistance, while thermal flux is the ratio of temperature difference and thermal resistance. Analogous quantities in both domains are summarized in Table 1.3.

Presence of a large variety of existing analysis tools and computation methods for electrical circuits promoted this analogy. Similar tools and techniques have

Table 1.3 Analogy between elements of a thermal and electrical system

Thermal metric	Unit	Electrical metric	Unit
Heat	J	Charge	C
Heat flow rate	W	Current	A
Heat capacity	J/cm^3K	Capacitance	F
R: Thermal resistivity	K/W	Resistance	Ω
Temperature	K	Voltage	V

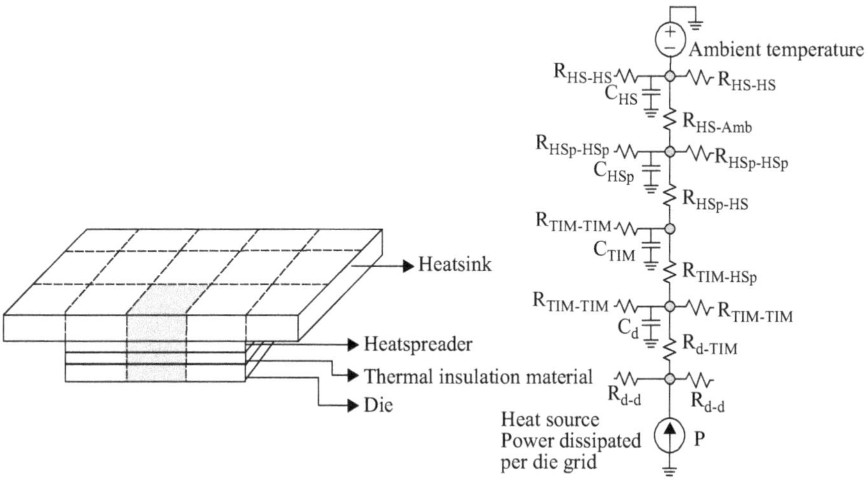

Figure 1.9 Compact thermal model of the chip package. A sub-network corresponding to one of the tiles in the partition (shaded region) is shown on the right. The complete model would have an interconnected grid of several sub-networks each corresponding to one tile

been widely adapted for thermal analysis and simulation tools, ranging from Finite Element Analysis to compact RC-circuit equivalent thermal models [59]. A widely adopted equivalent network representation is illustrated in Figure 1.9.

In this representation it is also common to divide the components into sub-blocks, where each grid point in the silicon layer represents an individual heat source. This circuit network model enables to incorporate variation in power densities across different components of a large-scale IC, such as the variation among different execution units in a processor core or across cores in a multicore chip. In this figure, a single grid location of the larger system is highlighted on the left and the equivalent circuit network is depicted on the right-hand side. Major heat transfer paths are denoted with thermal resistances. Thermal coupling also occurs between sub-blocks of the same layer in the form of a lateral heat path. The dominant heat dissipation path is along the vertical axis from the active layer of the die upwards towards the heatsink. Each layer (die, thermal insulation materials, heat spreader, heat sink) is represented with the corresponding thermal resistivity and capacitance, separated by nodes denoting the surface temperature of each layer. The aim of thermally robust packaging is to create a maximally strong thermal coupling (i.e., minimal total thermal resistance) between the die and the heatsink. The thermal resistivity (or conductivity) of a passive heatsink is represented as a single quantity in this abstract model. If this is indeed the actual deployment option for a system, then, the volume and material of the heatsink is optimized by the manufacturers to comply with the worst-case thermal output of the underlying chip. Thereby, it is ensured that the package is able to stabilize at a worst-case peak steady-state temperature complying with the maximum allowed rating of the chip manufacturer. It should be noted, however, that for high-performance system with active cooling options, this parameter is actually variable. Depending on the heat removal need of the system, additional external cooling mechanisms (air cooling fan rate, flow rate of cooling liquid, input current to a thermoelectric cooler (TEC)) can be ramped up to lend a lowered effective thermal resistivity for the heatsink. Ultimately, heat conduction paths terminate at the surface between the heatsink and ambient air.

1.2.2.1 Heat dissipation in vertical integration technologies

Other emerging technologies such as 2.5D interposer-based chips and 3D IC stacks exhibit thermal behavior governed by the same fundamental principles. However, the vertical integration introduces more pronounced thermal challenges and lead to additional design considerations, especially for multi-layer IC stacks. Figures 1.10 and 1.11 illustrate representative structures for a 2.5D and a 3D IC.

One of the prominent examples to commercial 2.5D chips is the Xilinx Virtex-7 FPGA device, which combines up to four FPGA dies attached to a silicon interposer [60]. Commercial realization of 3D IC stacks has been so far most successful in dynamic random access memory (DRAM) memory chips, such as the Micron

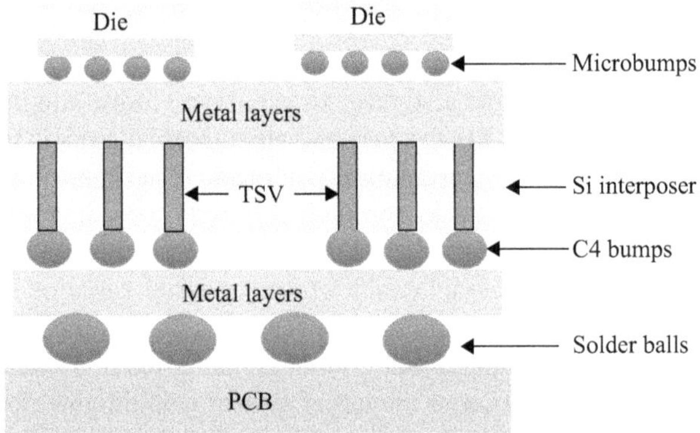

Figure 1.10 Depiction of a 2.5D IC: individual chips, the silicon interposer, and the PCB layers are interfaced with layers of metal layers and C4 and solder bumps. The TSVs embedded in the interposer layer route signals between metal layer interfaces

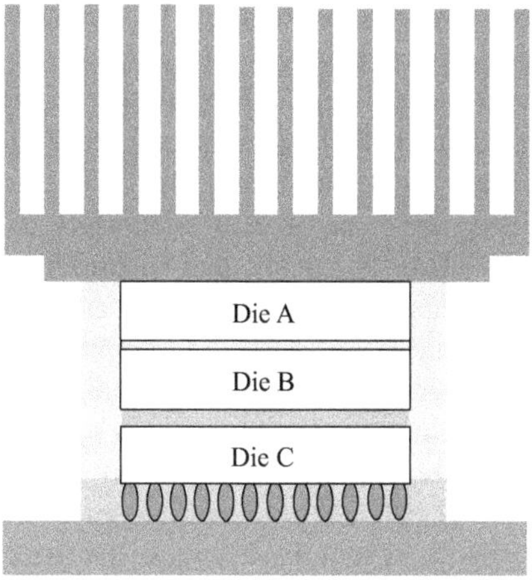

Figure 1.11 *Depiction of a 3D IC: individual die components constitute different tiers of the chip. TSVs facilitate inter-tier signal communication. This example illustrates a 3D IC with three tiers. The heatsink is attached to the topmost layer and the bottommost layer is interfaced with the PCB*

Hybrid Memory Cube [61]. Other 3D ICs have been developed for sensing applications, where monolithic MEMS are produced over digital IC layers [62] and mobile communication systems are another important target domain [63].

Layering dies vertically has the immediate consequence of those dies in the bottom tiers not having direct access to the heatsink surface. This creates a reduced effective heatsink surface for the overall IC and interior layers can only dissipate heat via conduction through the tiers above them. Also, there are important architectural and technology changes in chips that are deployed in vertical integration. For instance, ultra-high-speed communication among dies sharing the same interposer enables designers to allocate less local memory within the processor chip and move more memory off-chip onto the interposer. In turn, this intensifies the power density on the processor chips without the spreading (and cooling) effect of the memory arrays in the layout. Furthermore, dies used in stacked ICs are thinner and use different isolation materials, which result in poorer thermal conductivity and hamper their ability to dissipate heat [64].

The impact is less severe for 2.5D ICs that host multiple chips side by side, where each chip still has access to a dedicated heatsink. Still, bringing dissimilar chips of disparate power densities and heat output in close physical proximity might create thermal challenges, when they are conventionally not designed for such close encounter. For instance, placing high-performance multicore processor chips in close proximity to a stack of DRAM memory chips on an interposer might lead to an uneven thermal behavior of the memory stack. The side of the memory stack facing the processor cores with large heat output can exhibit higher temperature due to thermal coupling while the side facing away will be at a lower temperature. Such thermal imbalance within an IC can lead to variation in performance and reliability parameters (e.g., leakage and subsequent refresh dynamics in a DRAM, or uneven wear out) in chips subject to thermal aggressors nearby [65]. 3D ICs face more drastic thermal dissipation problems than their 2D and 2.5D counterparts [66, 67]. Stacking multiple dies into a 3D structure results in more total heat dissipation than the sum of each individual die's thermal output as standalone 2D structures. This is mainly due to the reduced net thermal conductivity of the 3D stack, which in turn causes heat being trapped within interior tiers of the stack. Naturally, the heat sink can only be in direct contact with the topmost die and all other tiers underneath need to dissipate heat via conduction towards the layers above them.

Figure 1.12 illustrates the thermal simulation of a combination of a 16-core processor chip with a 4-tier DRAM stack on an interposer. The processor die dissipates a total of 100 W. The DRAM stack dissipates 1 W per tier. The figure depicts the cross-sectional view on top and the thermal map of the bottommost die on the bottom. From the cross-sectional view shown in Figure 1.13 we observe that the coldest location of the chip is on the uppermost layer close to the side facing away from the processor core, while the side of the bottommost die facing the processor is hottest. The difference can be as large as 13.2°C across these locations in this case. This illustrates how bottom dies are in general at a disadvantage with respect to upper dies across a 3D stack and furthermore, when combined on a silicon interposer with thermally aggressive components, thermal stress can be further amplified.

Vertical integration technologies utilize various design optimizations to minimize heat buildup in local spot. An important decision to make is the ordering of the layers and assignment of functionality to tiers. Often, the most active dies hosting

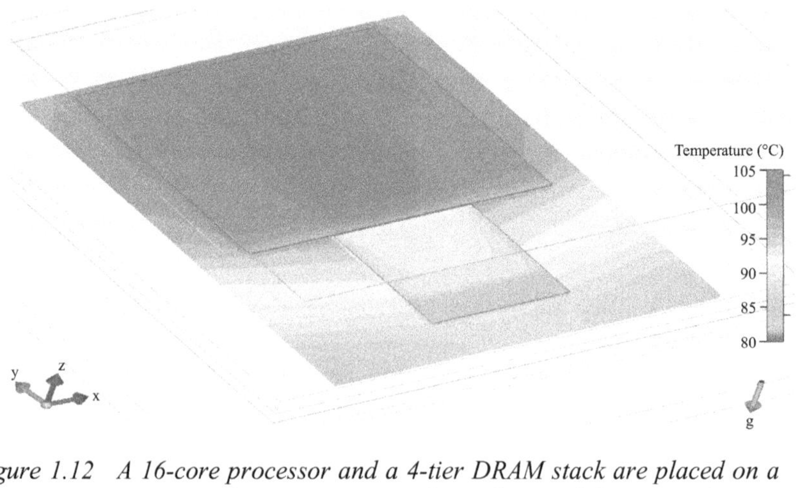

Figure 1.12 A 16-core processor and a 4-tier DRAM stack are placed on a common silicon interposer. Thermal map of the entire chip is illustrated under heavy processor workload (~100 W total)

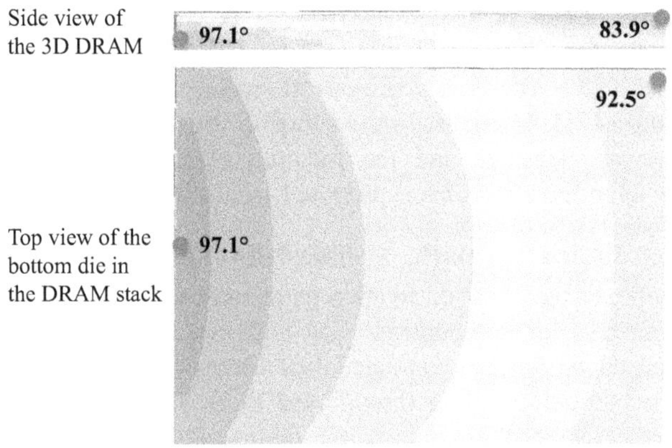

Figure 1.13 The thermal map of a 4-tier DRAM stack illustrates the impact of vertical integration onto lower layers of the stack. In addition, the influence of a thermally aggressive die onto its physically adjacent neighbors is noticeable

high-speed digital logic, such as processor cores, are placed on the topmost layers closest to the heatsink, while tiers with less power density and heat output, such as memory could be placed farther away from the heatsink. However, for certain make-ups of 3D ICs, widely heterogeneous functionality, that is, digital logic, MEMS, optical devices, might be present. Specific I/O needs and mechanical requirements would then dictate a particular ordering of the dies. In addition, design optimizations such as rotation and logic folding are considered when multiple dies of same functionality are combined. The die at a given tier can be positioned subject to a rotation with respect to the previous tier such that sub-blocks of high power density do not completely overlap in the vertical dimension across all layers.

While design optimizations could improve the power density and heat distribution to some extent, it may not remedy the problem entirely. Also, it may not be feasible to apply some of these solutions due to added design complexity. A structural solution to improve the conductivity of the thermal paths from the bottom layers towards the heatsink is the thermal through silicon via (TSV). Thermal TSVs are structurally equivalent to signal TSVs that are widely used in 3D IC processes to enable high-speed and high-bandwidth cross-layer signal communication. Similarly, 2.5D chips utilize TSVs to enable high-speed interconnect between the interposer and the package substrate. A TSV is essentially a vertical connection that penetrates through one or multiple dies and silicon substrates. Since TSVs are filled with conducting material (often metallic, such as copper and tungsten) they are an excellent means for creating additional heat transfer paths in the vertical direction. Thermal TSVs exploit this property and they are placed within vertical IC stacks as electrically isolated structures, purely for thermal coupling purposes. A similar solution for PCB-level heat management has long been used with thermal vias. Thermal TSVs can be placed within IC stacks in regular patterns or as concentrated in specific regions of high power density [68–70]. A common approach is to pre-allocate a certain fraction of chip area towards thermal TSVs and then place grids of thermal TSVs accordingly. Empirical studies often help to determine the optimal allocation [71]. Methods for customizing the placement according to the variation in power densities and the required effective thermal conductivity throughout the chip have also been proposed in academic research [72–74].

1.2.2.2 Heat dissipation paths at board-level and above

Beyond the chip package, heat dissipation paths exist that cause thermal coupling between various system components. For high-performance digital ICs, their internal heat production often dominates their thermal state with secondary contributions from thermal coupling through the PCB. There is a large variety of methods for packaging and bonding of the die to the external I/O and the PCB. Depending on the type of connections and materials used there will be thermal paths from the die's side facing away from the heatsink through the die pad, through the wire bonds or solder bumps into the PCB. Flip chip packaging using solder bumps provides a stronger thermal coupling between a die and the PCB, hence, it provides benefits for heat removal away from a high-performance chip.

At the board-level, the thermal state of certain components can be strongly influenced from board-level coupling, both through conduction and convection. A prominent example is the thermal interaction between memory chips sharing the same dual inline memory module (DIMM). Figure 1.14 presents a representative configuration of several components on the single board, including the cooling fan, processor chip, and memory DIMM. The thermal state of an individual DRAM chip would be influenced by multiple factors [65]. A chip's own internal power density is still a dominant factor. However, at the board-level relative placement of components will also dictate the efficiency of cooling that they will receive. This in turn, affects their individual thermal response. In the illustration of Figure 1.14, for example, the cool air reaches memory modules after having interacted with the hot processor chip. In this case, the downwind memory chips will be at the greatest disadvantage and

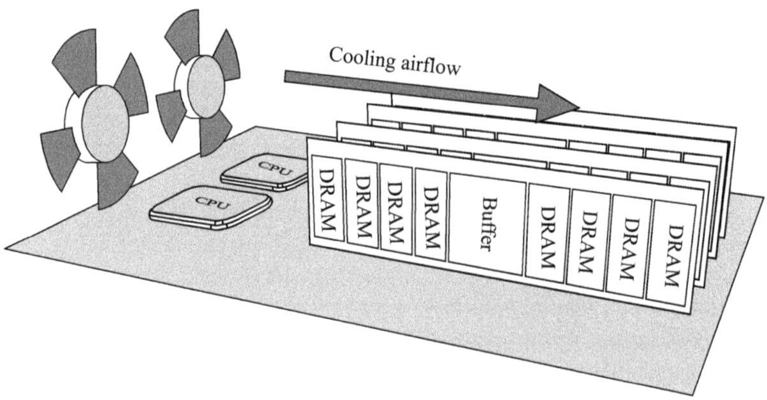

Figure 1.14 DIMMs placed alongside processor chips on a board

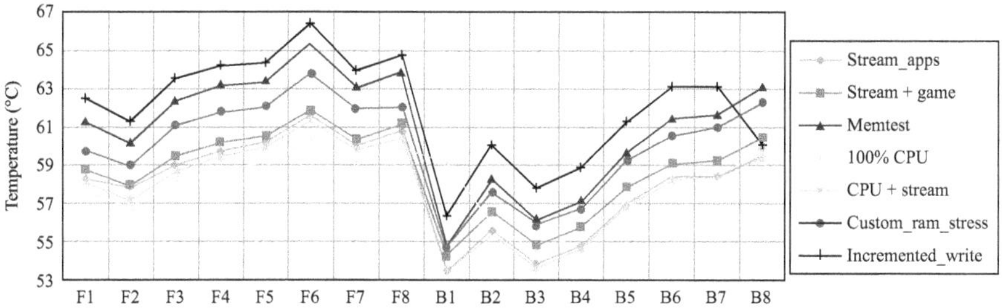

Figure 1.15 Variation in temperature across DRAM chips in a DIMM (front side chips labeled F1–F8, backside chips labeled B1–B8)

their thermal response under the same activity will be more severe than their upwind counterparts. Also, some DIMM configurations contain a buffer chip often placed at the center of a row of DRAM chips. This digital IC acts as a buffer between the memory controller and memory chips and it can facilitate control and other functionality such as error correction. These chips exhibit a significantly larger power density then the neighboring DRAM chips and impact the thermal state of the memory chips through thermal conduction. Some memory vendors also provide a unified heatsink spanning all of the memory and buffer chips. While this helps to spread the total heat and creates a larger surface area to interact with the cooling airflow, it also creates a direct thermal path from the hot buffer chip to the relatively colder memory chips, causing their temperatures to elevate. In summary, in addition to the lateral and vertical thermal paths within a given IC, it can encounter additional coupling due to sharing of a heatsink with other elements and due to being in close physical proximity on the board to a large heat source more active then itself. Figure 1.15 illustrates the variation in temperature in DRAM chips placed on two sides of a DIMM in presence of the abovementioned thermal paths.

References

[1] Flautner, K., *Keynote: The state of the future.* in *IEEE/ACM International Symposium of Microarchitecture*. 2010.

[2] Hamann, H.F., et al., Hotspot-limited microprocessors: Direct temperature and power distribution measurements. *IEEE Journal of Solid-State Circuits*, 2007. **42**(1): pp. 56–65.

[3] Petrini, F., et al., *What are the future trends in high-performance inter.-connects for parallel computers? [Panel 1].* in *High-Performance Interconnects, Symposium on*. 2004. IEEE Comput. Soc.

[4] White Paper: Managing Data Center Power and Cooling. 2007. Force 10 Networks, Inc.

[5] Pakbaznia, E. and M. Pedram, *Minimizing data center cooling and server power costs.* in *Proceedings of the 2009 ACM/IEEE International Symposium on Low Power Electronics and Design*. 2009. ACM.

[6] Long, J. and S.O. Memik, *A framework for optimizing thermoelectric active cooling systems.* in *Design Automation Conference (DAC), 2010 47th ACM/IEEE*. 2010. IEEE.

[7] Mahajan, R., C.-P. Chiu, and G. Chrysler, Cooling a microprocessor chip. *Proceedings of the IEEE*, 2006. **94**(8): pp. 1476–1486.

[8] Engelmann, C., *HPC resiliance research at Oak Ridge National Laboratory*. in *ER-OS: Scalable, Efficient, and Resilient Operating Systems*. 2010.

[9] Sato, K., et al., *Design and modeling of a non-blocking checkpointing system.* in *Proceedings of the International Conference on High Performance Computing, Networking, Storage and Analysis*. 2012. IEEE Computer Society Press.

[10] Das, B.P., et al., *Voltage and temperature scalable gate delay and slew models including intra-gate variations*. in *VLSI Design, 2008*. VLSID 2008. *21st International Conference on*. 2008.

[11] Wang, F., et al., On-chip bus thermal analysis and optimisation. *IET Computers & Digital Techniques*, 2007. **1**(5): pp. 590–599.

[12] Sundaresan, K. and N.R. Mahapatra, *An analysis of timing violations due to spatially distributed thermal effects in global wires*. in *Proceedings of the 44th Annual Design Automation Conference*. 2007. ACM.

[13] Lu, Z., et al., Interconnect lifetime prediction for reliability-aware systems. *Very Large Scale Integration (VLSI) Systems, IEEE Transactions on*, 2007. **15**(2): pp. 159–172.

[14] Xu, S., I. Benito, and W. Burleson, *Thermal impacts on NoC interconnects*. in *Networks-on-Chip, 2007. NOCS 2007. 1st International Symposium on*. 2007. IEEE.

[15] Chain, K., et al., A MOSFET electron mobility model of wide temperature range (77–400 K) for IC simulation. *Semiconductor Science and Technology*, 1997. **12**(4): p. 355.

[16] Beer, S. and R. Ginosar, Supply voltage and temperature variations in synchronization circuits. Technion, Haifa, Israel, Tech. Rep, 2013. **4562954**.

[17] Wolpert, D. and P. Ampadu, *Managing Temperature Effects in Nanoscale Adaptive Systems*. 2011. Springer Science & Business Media, New York, USA.

[18] Groeseneken, G., et al. Temperature dependence of threshold voltage in thin-film SOI MOSFETs. *Electron Device Letters, IEEE*, 1990. **11**(8): pp. 329–331.

[19] Filanovsky, I.M. and A. Allam, Mutual compensation of mobility and threshold voltage temperature effects with applications in CMOS circuits. *Circuits and Systems I: Fundamental Theory and Applications, IEEE Transactions on*, 2001. **48**(7): pp. 876–884.

[20] Calimera, A., et al., Temperature-insensitive dual-synthesis for nanometer CMOS technologies under inverse temperature dependence. *Very Large Scale Integration (VLSI) Systems, IEEE Transactions on*, 2010. **18**(11): pp. 1608–1620.

[21] Long, J. and S.O. Memik, *Inversed temperature dependence aware clock skew scheduling for sequential circuits*. in *Proceedings of the Conference on Design, Automation and Test in Europe*. 2010. European Design and Automation Association.
[22] Long, J., et al., *A self-adjusting clock tree architecture to cope with temperature variations*. in *Proceedings of the 2007 IEEE/ACM International Conference on Computer-Aided Design*. 2007. IEEE Press.
[23] Ni, M. and S.O. Memik, *Thermal-induced leakage power optimization by redundant resource allocation*. in *Proceedings of the 2006 IEEE/ACM International Conference on Computer-Aided Design*. 2006. ACM.
[24] Ni, M. and S.O. Memik, *Leakage power-aware clock skew scheduling: Converting stolen time into leakage power reduction*. in *Proceedings of the 45th Annual Design Automation Conference*. 2008. ACM.
[25] Ni, M. and S.O. Memik, *Self-heating-aware optimal wire sizing under Elmore delay model*. in *Proceedings of the Conference on Design, Automation and Test in Europe*. 2007. EDA Consortium.
[26] Yang, J.-S., et al., *TSV stress aware timing analysis with applications to 3D-IC layout optimization*. in *Proceedings of the 47th Design Automation Conference*. 2010. ACM.
[27] Sarangi, S., et al., *EVAL: Utilizing processors with variation-induced timing errors*. in *Microarchitecture, 2008. MICRO-41. 2008 41st IEEE/ACM International Symposium on*. 2008. IEEE.
[28] Chakraborty, A., et al., Thermal-aware clock tree design to increase timing reliability of embedded socs. *Circuits and Systems I: Regular Papers, IEEE Transactions on*, 2010. **57**(10): pp. 2741–2752.
[29] Srinivasan, J., et al., *The impact of technology scaling on lifetime reliability*. in *Dependable Systems and Networks, 2004 International Conference on*. 2004.
[30] Tu, K.N., Recent advances on electromigration in very-large-scale-integration of interconnects. *Journal of Applied Physics*, 2003. **94**(9): pp. 5451–5473.
[31] Thompson, C.V. and J.R. Lloyd, Electromigration and IC interconnects. *MRS Bulletin*, 1993. **18**(12): pp. 19–25.
[32] Hu, C.K., L. Gignac, and R. Rosenberg, Electromigration of Cu/low dielectric constant interconnects. *Microelectronics Reliability*, 2006. **46**(2): pp. 213–231.
[33] Black, J.R., Electromigration—A brief survey and some recent results. *Electron Devices, IEEE Transactions on*, 1969. **16**(4): pp. 338–347.
[34] Zhijian, L., et al., *Interconnect lifetime prediction under dynamic stress for reliability-aware design*. in *Proceedings of the 2004 IEEE/ACM International Conference on Computer-aided Design*. 2004. IEEE Computer Society.
[35] Banerjee, K., S.-C. Lin, and N. Srivastava, *Electrothermal engineering in the nanometer era: From devices and interconnects to circuits and systems*. in *Proceedings of the 2006 Asia and South Pacific Design Automation Conference*. 2006. IEEE Press.

[36] Lu, Z., et al., Improved thermal management with reliability banking. *Micro, IEEE*, 2005. **25**(6): pp. 40–49.
[37] Schroder, D.K., Negative bias temperature instability: What do we understand? *Microelectronics Reliability*, 2007. **47**(6): pp. 841–852.
[38] Li, X., et al., *Deep submicron cmos integrated circuit reliability simulation with spice*. in *Quality of Electronic Design, 2005. ISQED 2005. Sixth International Symposium on*. 2005. IEEE.
[39] Wu, E., et al., Interplay of voltage and temperature acceleration of oxide breakdown for ultra-thin gate oxides. *Solid-State Electronics*, 2002. **46**(11): pp. 1787–1798.
[40] Micron, T., *Technical Note: Uprating Semiconductors for High-Temperature Applications*. 2004. Intel Incorporation.
[41] Webb, W., Take the heat: Cool that hot embedded design-with processor performance skyrocketing and systems shrinking, designers are scrambling to find creative ways to live with the heat that today's. *Edn*, 2004. **49**(10): pp. 49–58.
[42] Application Note: Using Thermal Diodes in the PowerPC970MP® Processor. 2006, IBM.
[43] Dual-Core Intel® Xeon® Processor 5100 Series Thermal/Mechanical Design Guidelines. 2006. Intel Corporation.
[44] Floyd, M., et al., Adaptive energy-management features of the IBM POWER7 chip. *IBM Journal of Research and Development*, 2011. **55**(3): pp. 8:1–8:18.
[45] Rotem, E., et al., *Analysis of thermal monitor features of the Intel Pentium M processor*. in *TACS Workshop at ISCA-31*. 2004.
[46] Sanchez, H., et al., Thermal management system for high performance PowerPC(TM) microprocessors. IEEE Compcon 97, Proceedings, 1997: pp. 325–330.
[47] Arens, E.A. and H. Zhang, *The Skin's Role in Human Thermoregulation and Comfort*. 2006. Center for the Built Environment.
[48] Wasner, G.L. and J.A. Brock, Determinants of thermal pain thresholds in normal subjects. *Clinical Neurophysiology*, 2008. **119**(10): pp. 2389–2395.
[49] Ngak, C., New iPad hits whopping 116 degrees, report says. 21 March 2012, CBS News.
[50] Herrman, J., Why is my phone so hot? 21 September 2011, popularmechanics.com.
[51] Dingman, S., Hot new Sony Xperia Z2 smartphone's trouble with overheating. May 9, 2014, theglobeandmail.com.
[52] Berhe, M. and Y. Zhao, *Beyond Heat Spreading: Boosting Smartphone Performance through Phase Change Materials*. 2012. Henkel Electronic Materials, LLC.
[53] Troutman, R.R., VLSI limitations from drain-induced barrier lowering. *IEEE Journal of Solid-State Circuits*, 1979. **14**(2): pp. 383–391.

[54] Vassighi, A. and M. Sachdev, Thermal runaway in integrated circuits. *Device and Materials Reliability, IEEE Transactions on*, 2006. **6**(2): pp. 300–305.

[55] Tam, S., R.D. Limaye, and U.N. Desai, Clock generation and distribution for the 130-nm Itanium® 2 processor with 6-MB on-die L3 cache. *IEEE Journal of Solid-State Circuits*, 2004. **39**(4): pp. 636–642.

[56] Esmaeilzadeh, H., et al., Power challenges may end the multicore era. *Communications of the ACM*, 2013. **56**(2): pp. 93–102.

[57] Long, J., S.O. Memik, and M. Grayson, *Optimization of an on-chip active cooling system based on thin-film thermoelectric coolers*. in *Proceedings of the Conference on Design, Automation and Test in Europe*. 2010. European Design and Automation Association.

[58] Robertson, A.F. and D. Gross, An electrical-analog method for transient heat-flow analysis. *Journal of Research of the National Bureau of Standards*, 1958. **61**(2): pp. 105–115.

[59] Huang, W., et al., HotSpot: A compact thermal modeling methodology for early-stage VLSI design. *IEEE Transactions on Very Large Scale Integration (VLSI) Systems*, 2006. **14**(5): pp. 501–513.

[60] Dorsey, P., Xilinx stacked silicon interconnect technology delivers breakthrough fpga capacity, bandwidth, and power efficiency. Xilinx White Paper: Virtex-7 FPGAs, 2010: pp. 1–10.

[61] Jeddeloh, J. and B. Keeth, *Hybrid memory cube new DRAM architecture increases density and performance*. in *VLSI Technology (VLSIT), 2012 Symposium on*. 2012. IEEE.

[62] White Paper: The Advantages of Integrated MEMS to Enable the Internet of Moving Things. June 2014: mCube, Inc.

[63] Or-Bach, Z., Qualcomm Calls for Monolithic 3D IC. EETimes, June 6, 2014.

[64] Rinaldi, N., S. Russo, and V. d'Alessandro, *Thermal effects in thin silicon dies: Simulation and modelling*. in *Ultra-Thin Chip Technology and Applications*. 2011. Springer. pp. 287–308.

[65] Liu, S., et al., *Hardware/software techniques for DRAM thermal management*. in *High Performance Computer Architecture (HPCA), 2011 IEEE 17th International Symposium on*. 2011. IEEE.

[66] Jain, P., et al., *Thermal and power delivery challenges in 3D ICs*. in *Three Dimensional Integrated Circuit Design*. 2010. Springer, New York, USA. pp. 33–61.

[67] Tan, C.S., R.J. Gutmann, and L.R. Reif, *Wafer Level 3-D ICs Process Technology*. 2009. Springer Science & Business Media.

[68] Ni, M., et al., *Efficient design practices for thermal management of TSV based 3D IC system*. in *2010 International Symposium on Phycical Design (2010ISPD)*. 2010.

[69] Shi, B., A. Srivastava, and A. Bar-Cohen, *Hybrid 3D-IC cooling system using micro-fluidic cooling and thermal TSVs*. in *VLSI (ISVLSI), 2012 IEEE Computer Society Annual Symposium on*. 2012. IEEE.

[70] Sapatnekar, S.S., *Thermal via insertion and thermally aware routing in 3D ICs*. in *Three Dimensional Integrated Circuit Design*. 2010. Springer. pp. 145–160.

[71] Ni, M., et al., An analytical study on the role of thermal TSVs in a 3DIC chip stack. *in Proceedings of the Conference on Design, Automation and Test in Europe*. 2010.

[72] Cong, J., G. Luo, and Y. Shi, *Thermal-aware cell and through-silicon-via co-placement for 3D ICs*. in *Proceedings of the 48th Design Automation Conference*. 2011. ACM.

[73] Kim, D.H., R.O. Topaloglu, and S.K. Lim, *Block-level 3D IC design with through-silicon-via planning*. in *Design Automation Conference (ASP-DAC), 2012 17th Asia and South Pacific*. 2012. IEEE.

[74] Chen, Y., et al., Through silicon via aware design planning for thermally efficient 3-D integrated circuits. *Computer-Aided Design of Integrated Circuits and Systems, IEEE Transactions on*, 2013. **32**(9): pp. 1335–1346.

Chapter 2

On-chip temperature sensing

All on-chip temperature sensors operate on the common principle of generating an output voltage, current, or pulse that is correlated with the thermal state in the physical vicinity of the sensor. These devices are generally comprised of two parts, which we shall refer to as the back-end and the front-end. The back-end constitutes the actual sensing device in charge of producing a measurable voltage/current signal, while the front-end performs the conversion of this output signal into a digital representation, which system components utilize in downstream thermal management modules. Temperature sensors, which encapsulate both the back- and front-end components together on the actual spot of measurement, are also referred to as local sensors (or their measurements are called local sensing). In another realization, the back-end component might be placed at or near the spot of thermal event for monitoring, while the front-end component is located elsewhere on the chip or even off-chip. One front-end component might also be shared among multiple sensing devices in these realizations. A sensor implementation, where the two components are physically separated in this manner is referred to as a remote sensor. Finally, temperature sensors are classified as analog versus digital. We denote sensors that utilize doped p- and n-regions, a bipolar junction transistor (BJT) planted into the IC, resistive lines, or thermoelectric materials in their back-end structures, as analog sensors. Digital sensors utilize MOSFETs as the fundamental building block in various configurations (MOSFET as a diode junction, inverter delay lines, ring oscillators, leakage sensors, etc.) to construct the back-end.

On-chip integration of temperature sensors has occurred in two major phases, where early sensors were built using fully analog circuits largely exploiting the dependence of a forward-biased diode device's resistance on temperature. This was followed by the introduction of digital sensors in newer generations of chips, which are comprised of various forms of CMOS-based delay/leakage/time to temperature converters. Evolution of integrated on-chip temperature sensors have been largely driven by microprocessor manufacturers ever since they have started implementing thermal management infrastructures in their products. Diodes individually occupy less silicon, however, their accuracy is degraded by non-linearities over the operating range and with thermal and process variations. Often it is necessary to introduce additional circuits and calibration to improve stability and linearity. Digital sensors achieve better accuracy with less design and testing effort, at the expense of larger silicon area and power consumption. In this chapter, main types of analog and DTSs will be reviewed, along with their design, test, and production challenges. In addition, methods for inferring chip temperatures from various indirect measures and metrics will be discussed. Such methods can be used to supplement actual sensor data and create a finer grain thermal map of the underlying system than the granularity of the on-chip sensors' placement.

Before we review various types of on-chip sensor implementations, let us first comment on the general performance specifications provided by available design options and the expectations from these sensors in the field for different applications.

2.1 Operational conditions and performance specifications for on-chip temperature sensors

There are a number of performance metrics, which can be used to classify a temperature sensor. Table 2.1 summarizes the main parameters that are often used in practice and ranges of values for these parameters in commercial chips as well as reported from academic publications. Nanoscale on-chip integrated temperature sensors are subject to different thermal conditions and design constraints than their macroscale and discrete counterparts. While accuracy, linearity, and resolution might be equally important at either scale, on-chip sensors are expected to incur small area, low power consumption, high reproducibility, and robustness against process variations. Also, any calibration required for the on-chip sensors will add to the production cost to the host chip, while discrete sensors do not face this issue. Among on-chip sensors specifications also vary depending on the intended application. It is important to also note that, temperature sensors may serve different purposes on a chip. Highly accurate yet possibly more complex designs might be desirable for one chip manufacturer, while for another chip manufacturer smallest

Table 2.1 Common specification parameters defined for on-chip temperature sensors

Parameter	Unit	Range
Temperature range	°C	−120°C to 120°C (space) −70°C to 125°C (military/commercial) 20–125°C (processor, memory)
Accuracy	°C	±10°C (early sensors in processor chips) ±1°C (sensors in current processor chips) as low as ±0.15°C (academic design [1])
Resolution	°C	8–24 bits analog-to-digital converters (ADCs) 0.09–0.28°C (academic designs)
Sampling speed	Samples/s	2 samples/s–15M samples/s 50 samples/s (90 nm commercial processor [2])
Area	mm^2	0.006 mm^2 (22 nm processor chip [3]) 0.013–0.6 mm^2 (process nodes ranging from 22 nm to 350 nm)
Power	W	3 nW–2.88 mW

footprint with a larger error margin could be tolerable. For current on-chip monitoring needs the preference seems to be on achieving lower cost sensors and embedding as a fine-grained network as possible for highest hotspot coverage. Yet, emerging applications of thermal monitoring beyond DTM might require much higher accuracy and to detect smaller thermal variations, which in turn might be exploited to uniquely authenticate a specific chip or correlate thermal with process variation and other manufacturing aspects. In the following, we will review relevant sensor specifications and how they relate to the utility of the sensor with examples from state-of-the-art sensor implementations, both commercial and academic.

The largest range of temperature in which a sensor maintains its specified linearity and accuracy is referred to as the range of the sensor. This parameter is closely tied to the application intended for a given sensor. The range of temperatures commonly experienced by high-performance chips that employ thermal monitoring systems spans from room temperature ~20°C to ~125°C for commercial applications. In practice, the peak temperatures of these chips mostly remain under 90°C, with occasional higher thermal peaks (e.g., during Turbo Boost periods for state-of-the-art server-type processors). For vertically integrated 3D ICs these peaks have been reported to be exceeded and the worst-case operating temperatures at hotspots may run at and above 90°C more often than their 2D counterparts. Other higher stress applications, such as military and space applications, span a wider range from −50°C to 150°C. Various high-performance processor and DRAM chips are commonly rated by manufacturers to reliably operate up to ~125°C.

Within a given operating range, there are several important performance specifications for temperature sensors including accuracy, resolution, linearity, sampling speed, footprint, and power consumption. As we discuss the specifications of various sensors, we should carefully differentiate between those reported for sensors that are currently deployed in commercial products and those that resulted from academic studies. Sensor designs that have been deployed in the stat-of-the-art commercial chips have been tested for their reproducibility and robustness to variations at high volume production. On the other hand, other sensor designs in the academic literature have been subject to limited prototyping (ranging from a few to a dozen) and only simulation-based analysis has been performed at times. For this reason, best-case specifications reported for these sensors should be evaluated with this understanding. For example, current commercially deployed temperature sensors target a resolution of about 0.1°C and accuracy of less than 2°C, while specific sensors deployed in different products may be slightly above or below these values.

Accuracy/inaccuracy of a temperature sensor is broadly defined as the worst-case difference between what is measured and the real value of the measured quantity within the defined operation range. Another related sensor parameter is precision, which is defined as the smallest variation in the measured thermal quantity that a sensor is able to report correctly and consistently. The terms "high precision" and "high accuracy" are often used interchangeably in the literature concerning on-chip temperature sensors and in this context the term sensor accuracy is used to indicate the extent of accuracy in differentiating incremental temperature changes.

The concept of *thermal monitoring accuracy* is instead used to represent the gap between the reported temperature measurements of an on-chip thermal monitoring system and the actual temperature at the particular physical location. Often times in practice, this is further linked to the peak temperature present on a chip, since that is one of the most relevant conditions to carefully monitor in a chip. Then, monitoring accuracy translates into the level of fidelity achieved by a thermal monitoring system to track and report the true physical peak temperature of a chip.

As far the precision/accuracy of an individual sensor is concerned according to our earlier definition, the basic building block used to establish the temperature dependency is a major factor. For instance, diode- and BJT-based analog temperature sensors will be subject to device non-linearities and variations, which impact the junction/Base-Emitter voltage output produced proportional to temperature. This voltage output is the basis for the thermal measurement in analog sensors. Similarly, sensors that employ MOSFET transistor fundamentally generate a temperature-dependent response signal in the form of the Gate-to-Source voltage of a MOSFET. Intrinsic inaccuracies in the device parameters will introduce a discrepancy between the measured and actual quantities of these voltage outputs. Any process spread manifested in the output voltage and currents will also introduce inaccuracies. Process spread caused by issues such as dopant fluctuation, oxide thickness variation, threshold voltage variation, make MOSFET components more prone to inaccuracies. On the other hand, analog structures suffer more from noise and thermal dependency in their device responses. Also, the front-end structures of sensors, particularly ADCs are another source of inaccuracy. Both the back-end structures (BJT-based or MOSFTE-based) and the front-end ADC structures utilize one of more reference voltage and/or reference currents. Calibration often improves these inaccuracies and the operating range can be adjusted to reflect the consistent region of operation for a given sensor design. Nevertheless, any practical sensor covering a reasonably large range in on-chip applications will still exhibit some known amount of inaccuracy.

Resolution of a sensor relates to the smallest change in the measured quantity that the sensor can reliably differentiate and detect. A temperature sensor's resolution is largely influenced by the method used for converting the physical parameter (voltage, current, electric pulse, frequency) to a digital quantity. The front-end components, such as the ADC, counter used to perform this conversion, determine what precision can be achieved in representing the measurement. On the other hand, with increasing resolution the complexity of the conversion unit increases with associated area and power overheads. Therefore, in practice, often the least complex digitizer that is capable of achieving a certain resolution spec is preferred. For a given expected operating range R and a desired resolution r, the necessary number of digitized steps are given as

$$\textit{Total digitized steps} = \frac{R}{r} \tag{2.1}$$

In turn, the number of bits in the digital output of the sensor front-end, that is, ADC bitwidth, is given as

$$ADC\ bitwidth = log_2\left(\frac{R}{r}\right) \tag{2.2}$$

For example, a sensor that operates in a range of 100°C can provide a resolution of 0.1°C with a 10-bit ADC.

Linearity of a sensor is defined as its ability to operate according to a response function, which is a linear function of temperature. All sensors are expected to exhibit such a linear transfer function in one form or another. The maximum deviation from the linear response curve within the required operating range denotes the linearity parameter for a sensor. BJT-based back-ends leverage the close to linear relationship between the forward bias voltage at the junction diode of the transistor, MOSFET-based sensor back-ends aim to establish a linear relationship between the threshold voltages and Gate-to-Source voltages of carefully tuned transistors in a network, where threshold voltages in turn exhibit a dependency on temperature. Other sensors, such as thermocouples similarly exploit another linear relationship between an intrinsic metal property concerning charge carrier diffusion and temperature. Furthermore, front-end structures such as the ADCs exhibit non-linearities, which contribute to the overall sensor transfer function. Most practical on-chip sensor designs are expected to maintain linearity errors less than approximately 1°C.

Sampling speed is either defined by the maximum number of temperature readings collected from the sensor per time unit, or by the amount of time required to produce one measurement. This includes the activity of both the back-end and the front-end components. The sampling speed from a sensor will be relevant depending on the requirements of an application's thermal monitoring needs. Also, the need for the granularity of time resolution in sensing depends on the associated thermal time constants in the system. The speed at which temperature rises and falls in a chip has a direct influence on how often temperature should be measured. Thermal time constant for a chip is defined as the product of its effective thermal resistance and capacitance. It is a measure to the amount of time it takes for the system to settle into a steady-state temperature from an initial ambient temperature after a constant power density has been applied on it. For different chips of different material, packaging, and cooling make-ups, the thermal time constants will be different. In general, for high-performance processor chips thermal time constants are in the order of tens of milliseconds [4]. On-chip temperature sensors are expected to sample the chip's temperature at a compatible rate to be able to capture a fine grained thermal response. At peak performance and power bursts, temperature can rise by several degrees Celsius per millisecond interval [4, 5]. On the other hand, the higher the sampling rate of a sensor, the larger its power consumption will be. Similarly, other related infrastructure, such as interconnect to carry sensor readings to thermal management units will be proportionally more active.

Area of an on-chip temperature sensor influences the design overhead introduced by the sensor, as well as its coverage of the chip's thermal map, consequently the precision of the chip-level thermal monitoring. Sensors with smaller footprint can be embedded in a larger quantity on a chip making it possible to extract a more detailed thermal map at each sampling instance. The amount of coverage achieved throughout the total chip area is also referred to as the spatial resolution of the sensor. Each individual sensor will be able to accurately track the thermal events within a specific surrounding area. Thermal events beyond this range will be only tracked with larger errors, which may become completely unreliable and non-representative. Current sensors deployed in commercial chips exhibit different area overheads depending on their style of implementation. Sensors, which only deploy their analog back-end on the chip while maintaining the digital conversion front-end off-chip, usually take up the smallest chip real estate. This should be evaluated with the understanding that they will introduce other types of overheads such as routing of their analog output signals over longer distances and need for additional pins for them. Digital sensors with all components placed on-chip tend to be larger as far as silicon area is concerned. With the increasing number of sensors deployed on chips in newer generations, total area dedicated to sensors becomes non-negligible. It has been reported that Intel's Nehalem family of processors allocated more than 1% of their area for temperature sensors [6]. State-of-the-art sensors deployed in volume production commercial chips manufactured at 22 nm can be as small as 0.006 mm^2 [3].

Power consumption of temperature sensors is becoming increasingly important as the number of on-chip sensors is rising in newer generations of chips. Modern high-performance processors employ a variety of on-chip monitoring devices, including thermal sensors, delay sensors, current sensors for advanced power management and post manufacturing adaptation to process variation. Each such auxiliary feature introduces additional overheads and they sum up to a significant amount quickly. Therefore, designers need to carefully optimize both area and power consumption of the sensor devices in order to enable a largest scale network with highest coverage. Furthermore, thermal sensors today are not only relevant for the high power high-performance processor cores. Many emerging products, such as low power mobile chips also leverage thermal monitoring, since heat dissipation is becoming a challenge in densely packed electronics within small form factors, such as smartphones, tablets. In these systems, it is even more critical to design all on-chip components with extremely low power consumption.

In addition to the main parameters described above, other design aspects such as, self-heating, noise immunity, robustness against process variations, and reliability are prominent for temperature sensors just like any other on-chip component. We will touch upon some of these issues as we discuss design and manufacturing challenges for sensors in the later part of this chapter. Finally, we should note that some performance parameters mentioned above have interdependence. More resolution and higher accuracy often comes at the expense of larger area and power consumption. Similarly, higher sampling rates require more power. Ultimately, the overall cost of the thermal monitoring system also depends on the total quantity of sensors deployed in the system and how they are interfaced with the chip-wide thermal monitoring engines. The sensor interfaces with the larger monitoring system will be discussed separately in Section 2.6.

2.2 Analog sensors

Analog on-chip temperature sensors represent the influence of temperature as measurable variation on their output signals, such as voltage, current, and delay. These output signals are ultimately converted to an absolute digital measurement with ADCs. There are many alternative methods to designing these sensor back-end structures [7]. In the following we will review common approaches to both the back-end and front-end designs.

2.2.1 Thermal diode-based sensors

A common implementation of on-chip temperature sensors in practice uses the concept of bandgap circuits [8]. These sensors operate on the basic principle of monitoring the change in electrical resistance of a diode with varying temperature [9]. The voltage measured across the terminals of a diode that is biased with a constant current source has a direct dependency on the temperature. These devices can be fabricated on the chip using doped p- and n-regions or a parasitic PNP BJT available in CMOS processes.

The basic principle can be explained as follows. When a diode is first forced a fixed bias current I_{B1} and then another fixed bias current I_{B2}, then the forward bias voltage of the diode will change by ΔV_F. The absolute junction temperature T_D of the diode is related to I_{B1}, I_{B2}, and ΔV_F by

$$T_D = \frac{q\Delta V_F}{\eta K ln\left(\frac{I_{B1}}{I_{B2}}\right)} \tag{2.3}$$

where η is the ideality factor of the diode, K is Boltzmann's constant, and q equals the charge of an electron. In most practical sensor designs, the core building block of analog sensors is a pair of BJTs, where the Base-to-Emitter voltage (V_{BE}) of a BJT at a fixed bias current is correlated with temperature experienced by the device. The difference between the Base-to-Emitter voltages of a pair of BJTs, biased at I_{B1} and I_{B2}, respectively, are extracted from a differential circuit and used as a temperature-dependent output voltage ΔV_{BE}. This output is also referred to as a proportional to absolute temperature (PTAT) voltage V_{PTAT}. Finally, a temperature-independent reference voltage V_{REF} is generated by combining an amplified version of the differential voltage V_{PTAT} together with one of the temperature-dependent diode voltages. Figure 2.1 depicts the block-level diagram of an analog sensor using a bandgap reference circuit and thermal diodes made out of BJT junctions. Note that, while many bandgap reference-based analog sensors have these major building blocks in common, their specific implementation could differ. Hence, specifics of the interfaces between the major blocks and the circuit implementations of individual blocks are shown in a generic form.

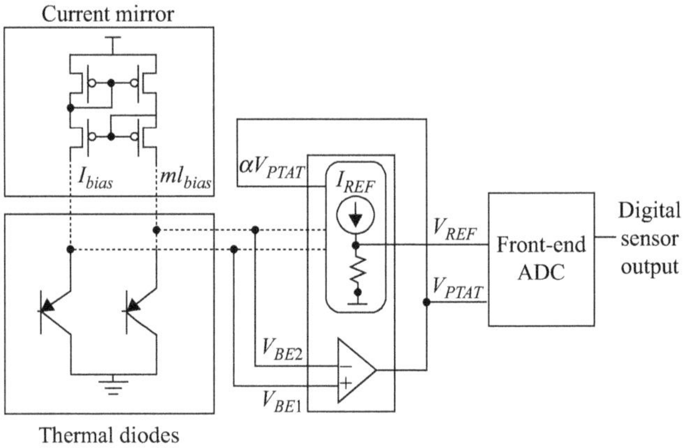

Figure 2.1 Illustration of main building blocks of a thermal diode-based bandgap reference circuit and its interfaces with the sensor front-end

A variety of sensors were designed based on this principle in different configurations [1, 8, 10–14]. In one of the realizations of BJT-based sensors deployed in several generations of Intel's processor chips from 90 nm to 32 nm designs [11], a temperature-independent reference voltage V_{REF} is also generated as a secondary output of the front-end component. This reference voltage is achieved by introducing a third current branch comprised of a diode and a resistor in series. The third branch is biased with a digitally controlled current source; a current steering digital-to-analog converter (DAC). ΔV_F exhibits a positive temperature coefficient, while V_{REF} obtained through a forward-biased diode exhibits a negative temperature coefficient. These two voltages are fed to a comparator in the back-end. The temperature measurement proceeds with scanning V_{REF} until the comparator indicates a match, in other words the tripping point where ΔV_F is greater than V_{REF}. The corresponding digital encoding of the scan parameter is sensed as the digital temperature measurement.

A sensor design utilizing diode structures was deployed in PowerPC cores [15]. This sensor utilizes two diodes constructed from p–n–p stack in the substrate to create a differential bias voltage stage, where the voltage output is a function of diode geometry and temperature. This bias stage is used in conjunction with a current mirror stage to generate two temperature-dependent current outputs. These current outputs are compared via a differential amplifier and ultimately processed with an ADC. Another bandgap circuit-based sensor was used in the IBM Power7 chips producing a voltage output [16–18].

An academic sensor design developed using similar principles proposed to generate V_{REF} from the pair of BJTs [10], where the V_{BE} output of one of the BJTs is added together with ΔV_F after an amplification to produce a bandgap reference voltage. The main trade-off points between the two abovementioned methods is that generation of V_{REF} from V_{BE} makes V_{REF} more susceptible to process variation. On the other hand, generation of an independent V_{REF} requires additional circuit elements. Other sensor designs utilize active materials to construct forward-biased p–n junctions [19].

One major shortcoming of using parasitic BJTs in CMOS chips is that they do not scale well with the rest of the CMOS logic and their forward bias voltage cannot be reduced at the same pace as the CMOS supply is reduced at newer process nodes. One proposal to address this issue is to utilize a Schottky metal-semiconductor diode as a replacement of the BJT device [20].

2.2.2 Resistive sensors

Thermal sensing by taking advantage of the change in resistivity of a conductor with respect to temperature is a well-known phenomenon. Thermistors are devices that have been built according to this principle as discrete macro-scale components. Thermistors can also be manufactured at nanoscale for on-chip applications. For example, IBM Power6™ processors introduced one of the first commercial uses of on-chip thermistors as an alternative to diode-based sensors [21]. These thermistors report temperature as the current resulting from applying a known voltage across their terminals. They are calibrated by measuring their resistance at known temperatures to derive the relationship between resistance and temperature. Their current output is sent off-chip to an ADC to generate the final digital temperature reading.

Other resistive temperature sensors have also been proposed in academic literature [22]. A ThermoWire utilizes a delay line comprised of segments of wires from the lower metal layers (M1, M2, M3) in conjunction with a time-to-digital converter unit. This implementation measures the time it takes for a pulse to travel through a resistive line and the front-end converts this propagation time to a digital temperature value. The thermal condition in these metal layers is expected to be tightly coupled with the underlying substrate's temperature. Also, this sensor is particularly well suited to measure the temperature of the metal layer itself, where the resistive line has been patterned. An alternative to this baseline structure is to create a dual delay line, where lower level metals are used for the first line and upper level metals are used for the second.

2.2.3 Thermocouples and thermopiles

The Seebeck effect arises when a temperature gradient across a conductor results in an electric field manifesting itself as a voltage difference between the endpoints of this gradient. The magnitude of this voltage per unit temperature difference (VK^{-1}) is an inherent property of a given conductor and it is quantified as its Seebeck coefficient. A temperature gradient across a conductor induces a diffusion of charge carriers until a counteracting electric field develops to sufficient strength counteract the carrier movement. For n-type and p-type semiconductors this phenomenon is dominated by the heightened energy and velocity of the electrons and holes, respectively, leading to a massive movement consistently towards the colder end of the gradient. Hence, all n-type semiconductors have a negative Seebeck coefficient while all p-type materials have a positive Seebeck coefficient.

For metals, other factors come into play to determine the net flux of carriers, such as the energy dependence of the mean free path scattering time of carriers [23]. Depending on these factors a net diffusion of carriers into either the hot or cold end might occur, hence, some metals have a positive Seebeck coefficient while others have a negative one. The bulk Seebeck coefficient values for a representative collection of metals are given in Table 2.2.

The temperature-induced voltage difference along a single strip of conductor could not be measured when a measurement wire of the same type is used to attach it to a voltmeter. The similar gradient across the measurement wire would cancel out the arising voltage difference. Hence, a junction with a dissimilar conductor needs to be constructed to measure the relative difference between the Seebeck voltages of the two conductors as illustrated in Figure 2.2. A voltage proportional to the junction temperature difference $T_H - T_C$ is thereby induced at the leads independent of the lead temperature T_0,

$$\Delta V_{HC} = \alpha(T_H - T_C) \tag{2.4}$$

Table 2.2 Absolute bulk Seebeck coefficients of select metals and alloys at 27°C. Most of the metals presented in the table are CMOS process-friendly

Metal	**Seebeck coefficient at 27°C (μV/K)**
Copper	+1.9
Aluminum	−1.5
Nickel	−10
Chromium	+14
Constantan	−30
Gold	+2
Bismuth	−65
Tungsten	+4

Figure 2.2 Conceptual drawing of a bimetallic thermocouple. ΔV_{HC} measures difference between T_H and T_C, independent of lead temperature and independent of the path of each lead according to equation 2.4

Here ΔV_{HC} represents the Seebeck voltage between the leads and α, the Seebeck coefficient. T_H and T_C are the temperatures at the hot and cold junctions, respectively. Thermocouples can therefore only measure a relative temperature as illustrated in (2.4) and hence, in order to function as a PTAT device, they need an additional reference measurement.

Thermocouples have long been used in macro-scale systems as temperature sensors [24–27]. Their consideration as on-chip temperature sensors for ICs is relatively more recent [28–35]. One of the first attempts at package-level integration used a macro-scale thin film bimetallic thermocouple array using copper-constantan pairing for package level sensing [30]. One of the early CMOS compatible on-chip sensor designs proposed using a metal/p+ diffusion or a metal/polysilicon couple [28]. Photolithographically patterned thin film bimetallic thermocouple sensors were proposed more recently starting with a conceptual design [35] followed by fabrication and characterizations results [34, 36]. This sensor is using Chromium–Nickel (Cr–Ni) pair and it can be directly integrated into the metal layers of an IC due to the choice of CMOS compatible metals. A conceptual illustration of a recent proposal for integration of a thermocouple array layered over an IC is shown in Figure 2.3. These results were followed by another demonstration of a (Cr–Ni) thermocouple sensor array [29, 33].

Figure 2.4 depicts the characterization of a thin film thermocouple temperature sensor deposited on top P-doped boron silicon wafers of 380 um thickness [31]. A 1 µm silicon dioxide layer separates the sensors from a resistive heater network layered underneath for characterization. The thermocouples are made of copper and

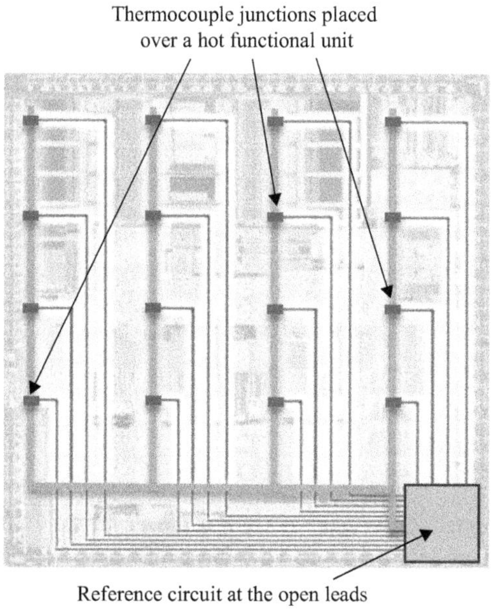

Figure 2.3 Conceptual illustration of integration of a thermocouple array

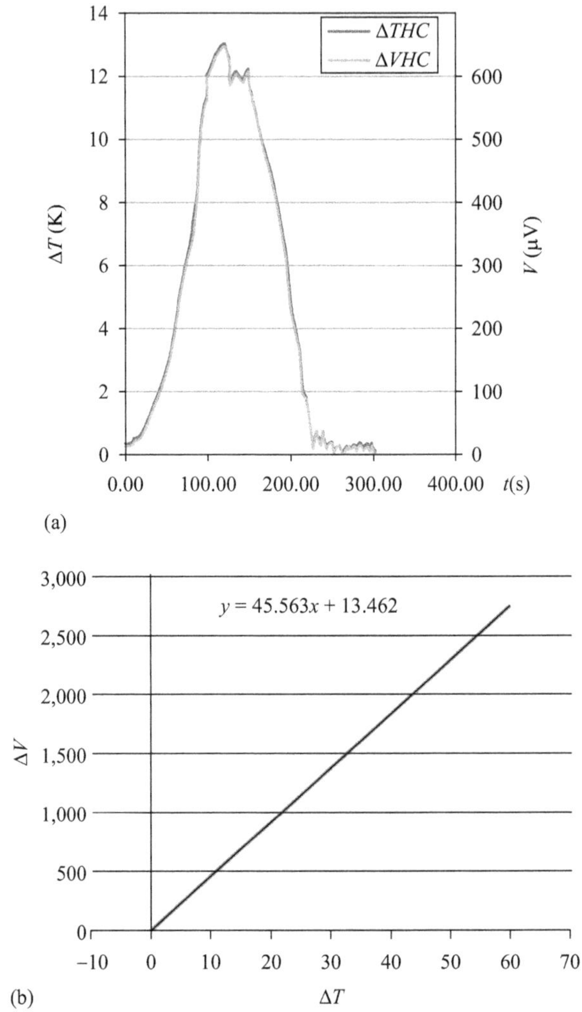

Figure 2.4 Characterization of a photolithographically layered thin film bimetallic thermocouple sensor: (a) transient behavior of the thermocouple output voltage and the temperature gradient reported by the thermistors over time as a heat pulse is applied, (b) parametric plot of the relationship between thermocouple output voltage and the temperature gradients experienced by the thermocouple. The slope is the Seebeck coefficient of this physical device

constantan. The copper leg of the thermocouple is about 140 nm thick and the constantan leg is about 250 nm thick. The characterization is performed by creating a thermal gradient between the joint and open leads of the thermocouple using the resistive heater element layered underneath. A heat pulse is created for this purpose. Calibrated thermistors attached on the backside of the sample at the two ends of the gradient are used to characterize the thermocouple sensor. Figure 2.4(a) depicts the response of the thermocouple output voltage over time as the heater is activated to create a heat pulse. On the same plot the temperature difference reported by the thermistors is also shown. Figure 2.4(b) presents the parametric plot of the relationship between the thermocouple's output voltage and the temperature gradient between its two ends. Within this temperature range (up to 60°C above room temperature), which is consistent with the practical operating range of modern chips, this sensor can be calibrated to operate in a linear range.

While promising results have been reported on the design and fabrication of thermocouples as thin films and their chip-level integration [29, 34], there are no complete architectures and deployment in commercial products yet. A recent study provided some preliminary guidelines on the integration of a network of thermocouple sensors in to 3D IC [31]. On the other hand, thin film bimetallic thermocouples present unique advantages for future on-chip sensing systems. Therefore, a growing interest in leveraging thermoelectric materials in sensing and other avenues of thermal management for CMOS chips is expected. ΔV_{HC} output of a thermocouple is independent of the width and length of the metal strips, as well as the possible variations in the thermal profile that may exist along the entire length of the strips. Hence, the strips can be formed in any shape. The only component that needs to be placed in the active area is the reference point sensor, which can be built using any existing sensor design paradigm. One parameter, which has an impact on ΔV_{HC} is the thickness of the metal layers, which becomes significant when the thickness of the film approaches the mean-free path of the electrons in the material. This parameter can be optimized for a given process technology and metals of choice.

Figure 2.5 illustrates components of a potential implementation of the basic sensor structure [31]. Metal strips A and B reside in two different metal layers and comprise the thermocouple. The location of the thermocouple junction where the two metal strips join is called the *sensing point*, and the location on the die corresponding to the hotspot of interest is called the *monitor point*. In order for the sensing point temperature to accurately track the monitor point, thermal coupling between these two points can be established. For this purpose, a thermal shortcut

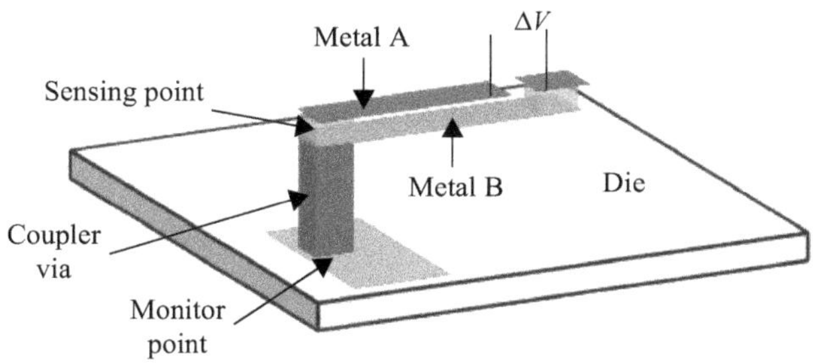

Figure 2.5 Illustration of a thermocouple attached to a sensing point and reading an output of ΔV

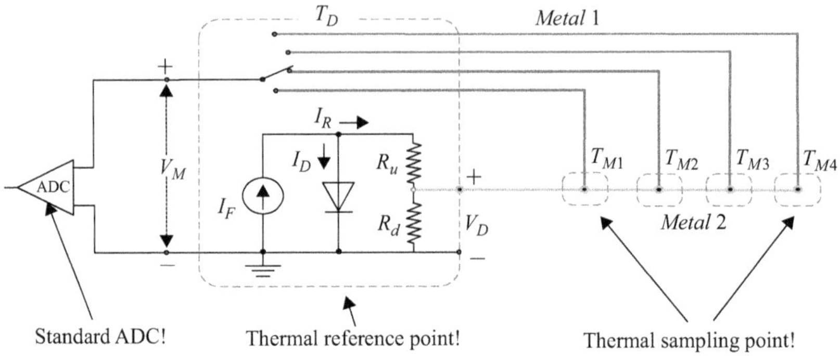

Figure 2.6 Circuit diagram of the proposed bimetallic integrated on-chip thermocouple array temperature sensor

between the sensing and monitor points can be created using a *coupler via*. The purpose of this via is similar to thermal vias used in ICs for heat management, that is, to create heat conduction paths. However, unlike thermal vias, which function as paths to move heat to the heatsink, sensor coupler vias thermally connect two specific points. Coupler vias are electrically isolated and serve no electrical function.

A larger array can be built by replicating this structure as in Figure 2.6. Each thermocouple generates a voltage output at its open leads proportional to the temperature difference between the remote junction and the common reference point. All open leads can be consolidated in close physical proximity to the reference circuit. To deduce the absolute temperatures, a single absolute-temperature sensor is placed at the common reference point. The placement location for the reference point could be any available free space at the outer edge of the chip, where thermal fluctuation is minimal and temperature remains close to the ambient, for which the total sensing system can be calibrated. Figure 2.6 illustrates a possible implementation of the reference circuitry, which employs a switch to combine multiple thermocouple outputs. Thereby, one reference circuit can serve a number of sensing locations throughout the chip. Depending on the size of the sensor array, one or more reference circuits can serve the entire array. Finally, the output of the reference circuit will be converted to a digital representation using some form of ADC. Since the sampling rates for thermal sensing applications are relatively low, sharing the thermocouple output switch, the reference thermal diode, and the ADC would be feasible.

2.2.4 Other types of analog sensors

In addition to the thermal diodes and metallic structures there are proposals to exploit other properties of silicon material or on-chip active devices to create temperature sensors from them. One such proposal is to exploit the dependence of silicon's diffusivity on temperature [37, 38]. A heater element and a thermopile are placed at a known distance, *s*, apart. A thermopile is essentially a collection of serially connected thermocouples. In this case, the thermocouples are made with the coupling of p+-diffusion regions and aluminum. A heat pulse created by the heater will experience a filtering process through the substrate subject to its temperature-dependent diffusivity. The resulting response of the thermopile is then processed by a readout circuit.

2.3 Digital sensors

Temperature sensors that are built exclusively using CMOS logic are generally referred to as digital sensors. Within this family of sensors there is a larger variety of design styles, converting the impact of temperature on resistivity, delay, leakage, etc. of CMOS logic into a digital representation of temperature.

2.3.1 MOSFET-based voltage/current output sensors

A possible approach is derived from the same basic principle as the BJT-based analog sensors, only replacing the BJT components with MOSFETs operating in the subthreshold region [39–42]. Reference voltages in the differential circuit would be derived from the Gate-to-Source voltage V_{GS} of the MOSFET instead of the V_{BE} of the BJT. The difference between the Gate-to-Source voltages of two MOSFETs follows similar dependency on the temperature involving the same set of parameters (shown in (2.1)) as in the case of BJT-based sensors. Various alternatives utilizing MOSFETs to construct temperature-dependent voltage [43–45] or current outputs [46, 47] have been proposed. Furthermore, recent design efforts have been targeted towards enhancing these sensors with self-compensation for process variation. One example is a CMOS sensor enhanced with a switched capacitor amplifier and sample and hold circuit to subtract common-mode process induced noise from the sensor output [48].

The abovementioned designs are examples of academic studies. Arguably the biggest difference between academic designs and commercial designs is the scale of validation. Several academic sensor design proposals are more sophisticated in various dimensions (process insensitivity, low energy consumption, etc.), however, they have not been tested under high volume production. Most designs have been fabricated with a small number (<10) of samples for characterization. On the other hand, there are also CMOS digital sensor designs of similar architectures that have been manufactured and tested at large scale with commercial processes. A CMOS digital sensor designed and tested at Intel's 32 nm processing node [3] provides a compact all digital design with significant area reduction in comparison to the BJT-based sensors deployed in earlier generations of processors [11]. This sensor, referred to as a Thermal Probe, is calibrated to operate at a limited yet higher temperature range (80–100°C) near the triggering point of the DTM intervention unit. It is intended for deployment at targeted hotspots while still maintaining a small number of the original sensors on the chip with higher area overhead yet wider operation range. The Thermal Probe circuit is comprised of three stages. The first stage is a voltage reference circuit in charge of generating a reference voltage slightly above the threshold voltage V_{th} of this process. In the second stage a current reference circuit is constructed by placing a MOSFET into linear operation range.

This is achieved by forcing the drain voltage of a MOSFET to be equal to the reference voltage with a negative feedback loop through an amplifier. At the same time the gate bias of this MOSFET is maintained at a higher level. The resulting current behavior of the MOSFET in linear region is a parameter of material properties and operational parameters:

$$I_{linear} = \mu C_{ox} \frac{W}{L} \left((V_{GS} - V_{th}) V_{DS} - \frac{V_{DS}^2}{2} \right) \tag{2.5}$$

As shown in (2.5), among the parameters defining I_{linear}, the carrier mobility μ and the threshold voltage V_{th} are those that have a direct (negative) dependency on temperature. This achieves a temperature-dependent current source. In the last and third stage, this reference current is replenished with a current mirror and used to charge a capacitor. The integrated output is continuously compared to the reference voltage and used to generate a pulse with its frequency coupled to temperature.

Another proposal utilizes an interface-trap charge pump [49]. The sensor involves a two-stage process of current to frequency and frequency to current conversions with a feedback loop. The interface-trap charge pump is essentially a frequency-controlled current source. It is combined with another temperature-dependent current source to derive a differential current, which is converted to frequency. This frequency in turn controls the interface-trap charge pump through the feedback path. Through this feedback, a linear relationship between the output frequency and temperature-dependent current is established.

2.3.2 Time-to-temperature (delay-line) sensors

Another popular approach to DTS design is to create a temperature-dependent delay line. There are various alternatives to this basic idea [50–54]. One of them is to create two parallel signal paths, one comprised of an even number of inverters constituting a thermally sensitive path and the other created through compensation techniques to ensure a thermally insensitive path [50]. For this method to function, it is indeed sufficient to create a measurably large difference between the temperature sensitivities of the two paths. The second path does not even need to be perfectly insensitive. Figure 2.7 illustrates such a sensor's structure. Both lines are activated with a periodic square wave *Enable* signal. The difference between the

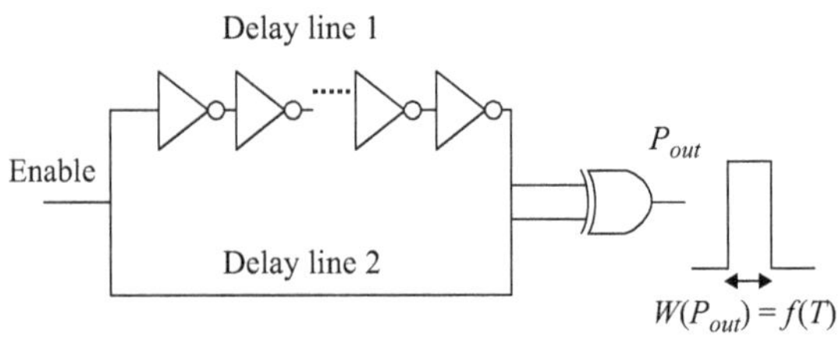

Figure 2.7 Diagram of a delay line sensor: the difference in the propagation time of the same input pulse through two delay lines is interpreted as an output pulse with its width proportional to temperature

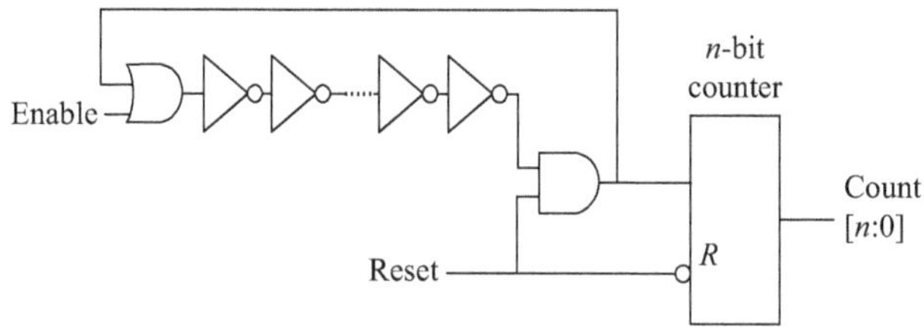

Figure 2.8 Diagram of a circular time to temperature sensor: the rate of oscillation frequency is captured with a counter. This quantity now correlates with temperature

arrival times of the signals at the end of either path is represented with a pulse P_{out} of a width proportional to temperature created at the output of an XOR gate.

A similar pulse-to-temperature measurement is achieved by observing the variation in a clock source under thermal effects. This sensor uses a pair of delay locked loops in conjunction with a circular inverter chain to generate a temperature-dependent and a thermally stable delay reference, respectively [53].

Ring oscillators have also been used to construct time-to-temperature sensors. While different realizations exist [55–64], a common idea is to convert the temperature-dependent frequency of a ring oscillator into a digital temperature reading utilizing a counter or a pulse generator similar to the abovementioned delay line sensors (as shown in Figure 2.8). If the output of the sensor is a pulse, then, an additional front-end block will be used to convert this pulse to a digital representation of temperature. If the cycle count of the ring oscillator is collected for a fixed period of time dictated by the *Enable* signal, the counter's output can be used directly as a temperature metric subject to calibration.

2.3.3 Leakage-based temperature sensors

Leakage sensors exploit the relationship between leakage currents in CMOS devices and temperature in various shapes and forms. Typically they require smaller footprint and energy with respect to their analog and other digital counterparts. On the other hand, efforts to provide self-tuning or calibration to achieve robustness against process variation and noise may be more pronounced. Hence, a trade-off point to consider would be to embed a larger number of small sized low power leakage-based sensors to operate in a more restricted thermal range versus larger yet more stable alternative sensor designs. Leakage sensors leverage the dependency of a MOSFET's subthreshold leakage current on temperature as depicted in (2.4), where the subthreshold current depends exponentially on temperature T, and also other parameters, such as an electron's charge q, Boltzman constant K_B, and device parameters represented with k, a, and finally the threshold voltage V_{th}.

$$I_{sub} = k\, e^{-qV_{th}(aK_BT)} \tag{2.6}$$

One early implementation utilized the decay time of charge stored in a 4T memory cell [65] as a measure of temperature. This design uses a comparator (one

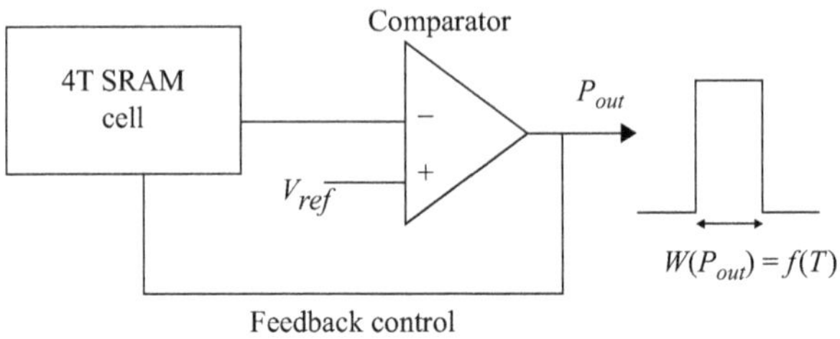

Figure 2.9 Diagram of a leakage sensor capturing the temperature-dependent rate of leakage in an SRAM cell: the rate of leakage from a 4T-memory cell is captured with a comparator and converted into an output pulse with its width proportional to the rate of leakage, which in turn correlates with temperature

input fixed at a low 90 mV and the other connected to the cell's output) and toggle flip flop (T-FF) in a feedback path to connect to the wordline control of the cell. A fully charged memory cell's output causes the comparator's output to be high, activating the T-FF to disable further charging of the cell. From that point onwards the cell's stored charge leaks according to the subthreshold leakage mechanism of its internal MOSFETs. When the cell's output drops below a threshold, the comparator's output drops to low, engaging the T-FF and, in turn, the wordline to restart a new charging epoch. The high pulse occurring in the meanwhile at the buffered output of the comparator constitutes the sensor back-end's output. The width of this pulse is associated with a digital temperature reading by the front-end. This sensor is illustrated with a block diagram in Figure 2.9.

Another sensor implementation correlates temperature with the time for a capacitor to discharge through the subthreshold leakage current path of a MOSFET [66]. The capacitor is combined with a pair of MOSFETs, where the pull-up PMOS in this pair serves to charge the capacitor upon activation with a high pulse, while the pull-down NMOS provides the discharge path. The voltage across the capacitor drives an inverter. The output of this inverter will then exhibit a low pulse where the width of the pulse has an exponential relation to temperature. This pulse will be converted to a digital reading via a front-end structure.

An enhancement to the sensor employing a discharging capacitor was to utilize the ratio between two measurements of a capacitor's discharge rate at two different reference voltages to account for process variations [67, 68]. This sensor also enhances the charging circuitry for the capacitor with an additional charge pump. An illustration of the leakage sensor tracking the discharge time is presented in Figure 2.10 [67].

2.4 Sensor front-end

Although we refer to thermal diode structures and thermocouples as analog sensors, it is with the understanding that their voltage or current output will be eventually converted into a digital representation. Ultimately, a digital value representing the

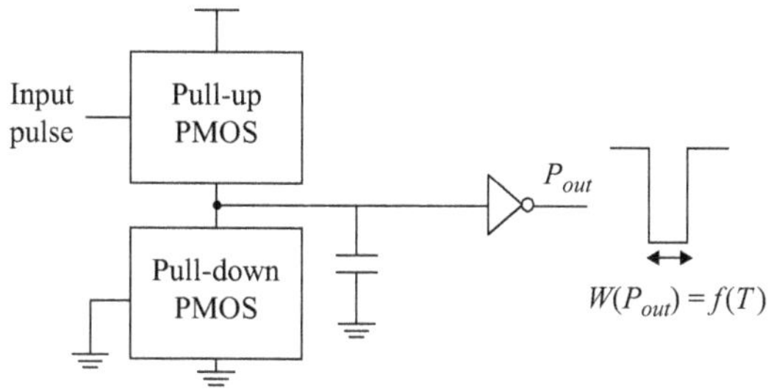

Figure 2.10 Diagram of a leakage sensor capturing the temperature-dependent discharge time of a capacitor: the rate of leakage from capacitor is captured with a pull-up/pull-down network and converted into an output pulse with its width correlating with temperature

measurement is provided to all subsequent thermal management units on the chip. Often, this digital output is compared against a threshold value stored in a register, representing a critical threshold value. If the comparison against this pre-stored value indicates that the threshold has been exceeded, then hardware and/or system software is notified to take thermal management actions.

Most implementations utilize various ADCs to represent the sensor output as a digital absolute quantity. There are three main types of converters commonly used for temperature sensing back-end: Sigma-Delta ADCs, successive approximation register (SAR) ADCs, and pulse-to-digital converters (PTDCs). Note that, not only sensors with diode or BJT front-ends, but also those with MOSFET-based front-ends similarly require converters from voltage/current to digital output. Some digital sensors also employ time-to-digital converters that directly digitize the width of an output pulse or the frequency of a periodic output signal.

In some implementations of DTM, the raw output from the sensor can be further refined through additional computational methods. For example, the IBM Power7 processor chip utilizes an analog sensor coupled with an ADC front-end. Then, this digital temperature reading is further refined by a dedicated on-chip microcontroller in software to fit the digital output of the ADC to a pre-programmed polynomial curve. This algorithm is programmed into the thermal monitoring scheme and the curve is customized for each individual sensor on the chip to account for the post manufacturing testing and calibration characteristics of each chip and sensor.

In the following we will review the basic operating principles of these front-end structures.

2.4.1 *Sigma-Delta ADCs for temperature sensors*

Sigma-Delta ADCs are suitable for thermal monitoring applications, where temperature sensors do not generate high frequency signals, but high resolution can be provided at reasonable area and power cost. This converter employs a Sigma-Delta Modulator [69] to convert the input voltage to a representation called the digital bitstream. Sigma-Delta Modulators employ a sampling rate that is faster than the

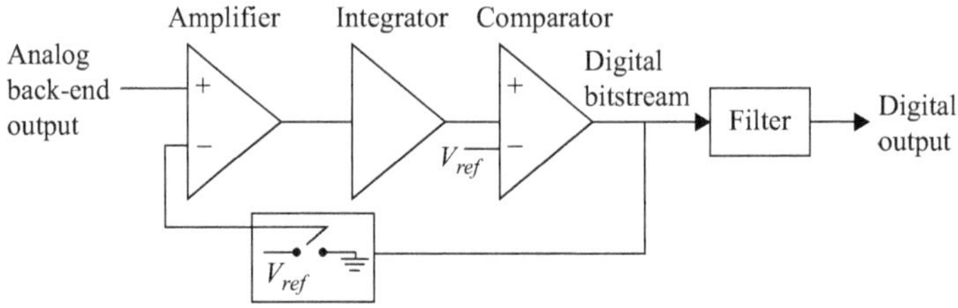

Figure 2.11 Block diagram of a first-order Sigma-Delta Modulator ADC

Nyquist criterion and hence, they are called to perform oversampling. A first-order modulator is illustrated in Figure 2.11. The modulation stage uses a differential amplifier, an integrator and a comparator to oversample a slowly changing signal. Higher order converters repeat this stage multiple times in sequence. The input voltage is continuously integrated and each time the threshold of the comparator has been reached, it generates a single bit output. At the same time the reference voltage is subtracted from the input voltage by the differential amplifier through a feedback path to reset the integrator's output. The comparator's output bit will remain high for the duration of time that the integrator transitions from the high threshold of the comparator to its low threshold. The comparator will then output a zero. Through this basic operation principle the modulator generates an over-sampled bitstream output. The bitstream is finally processed with an averaging filter and a decimating filter to generate the digital representation.

Discrete off-chip sensors can readily interface with another discrete ADC IC. However, on-chip temperature sensors designs are developed inclusive of the conversion circuitry and often involve advanced features to enhance the stability and process independence of the sensor [10, 69–71].

2.4.2 SAR ADCs for temperature sensors

An n-bit SAR ADC utilizes a comparator, which receives the latched input voltage V_{IN} and the output of an n-bit DAC. The DAC is fed from the SAR register, which is usually initialized with the value of the middle of the pre-determined operating temperature range. This SAR register counts by trying all values of bits starting with the most-significant bit (MSB) and finishing at the least-significant bit. The corresponding reference voltage is generated by the DAC and it is compared against the input voltage. If V_{IN} is larger then, the comparator's output is high leading to the MSB in the register to be set high, otherwise the MSB will be set to zero. In a number of successive steps, the register monitors the comparator's output to see if the binary count is less than or greater than the analog signal input, adjusting the bit values accordingly. The comparator's output is also used to drive a control circuit co-located with the SAR register, which then manipulates the next significant bit location in the register and repeats the comparison of V_{IN} with the new V_{REF}. Through this iterative process the input voltage value will be converted to an n-bit digital representation in n-cycles, hence, the SAR ADC usually generates one bit of its output at a time. Figure 2.12 illustrates the main components

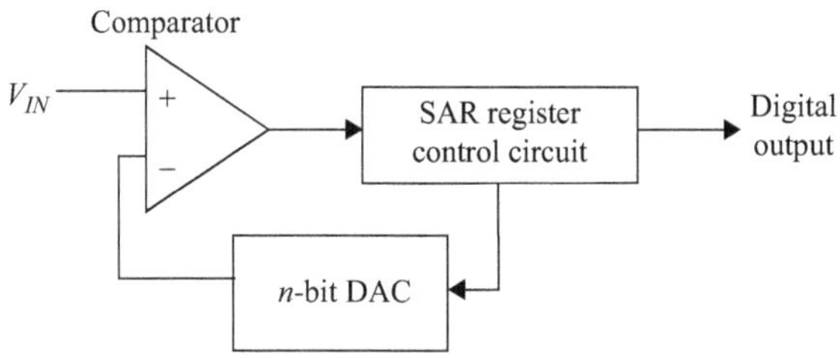

Figure 2.12 Feedback loop of the SAR-based analog to digital (A/D) conversion front-end

of the SAR ADC. On-chip implementations of SAR ADCs integrated with analog sensor back-ends have been proposed in literature [51, 52, 72].

2.4.3 PTDCs for temperature sensors

Another type of converter proposed to function as a front-end for analog temperature sensors is the PTDC. This converter is suitable to interface sensors that generate a current output. The structure employs two capacitors that are pre-charged up to a reference voltage immediately upon initiation of conversion. Then, the temperature-dependent current output I_{IN} and a reference current (similar to V_{REF}) start discharging these capacitors. Buffers driven by the output voltage of these capacitors will each produce a switching signal upon complete discharging of the respective capacitor. The timing of this will be correlated with the difference in the discharging time, which in turn is correlated with the temperature-dependent current. The difference in timing will create a pulse, whose width is measured in counts of cycles and converted to a temperature value.

An alternative that does not involve analog components captures the pulse within a circular structure (similar to the structure depicted in Figure 2.8) tuned to shrink the pulse by a fixed amount during each iteration of circulation. This structure is attached to a counter, where the number of roundtrips made by the pulse before dying off through the circular delay line is recorded [50, 73, 74].

2.5 Design challenges for sensors

Major design and manufacturing challenges for temperature sensors involve mitigation of non-idealities, achieving robustness against process, voltage, and thermal variations, and calibration. In addition, minimizing the footprint and power consumptions of the sensors are important targets to achieve, since they directly correlate with the level of granularity of the thermal monitoring system and the ease of placement of a sensor within densely packed and thermally significant regions of the layout.

2.5.1 Ideality and linearity

For BJT-based sensors, one of the most prominent challenges in design of analog sensors is to achieve desired accuracy in the presence of device non-idealities and process variation. First, the accuracy of an analog sensor depends on the precision of the constant current source used to bias the BJT structure. This subcomponent involves various regulation circuitry and stabilization enhancements in order to create linear and robust reference currents. One of the main challenges encountered by the design community has been to devise such enhancements to the voltage and current regulation circuits to achieve the desired stability at low area and power overhead.

Differential signals generated within the circuitry are by nature highly reproducible with as little as ±0.1°C variation [75], since variation factors tend to cancel each other out. However, the bandgap reference voltage derived from the Base-to-Emitter voltage of one of the BJTs may experience inaccuracies, due to the fact that in practice the relationship of V_{BE} and temperature is not perfectly linear due to non-ideality of the diode junction. Figure 2.13 illustrates this phenomenon. Figure 2.13(a) depicts the ideal behavior if perfectly linear temperature dependence is assumed for the forward bias voltage of the diodes. Figure 2.13(b) shows the non-ideal behavior of V_{BE} as a function of temperature, which in turn introduces a curvature to both the behavior of V_{BE} and V_{REF}. Sensors that utilize the bandgap reference voltage as an input to their ADC front-end will suffer from this non-ideality.

Methods to combat the curvature to achieve better linearization over a larger range have been proposed with various enhancements to the basic sensor structure [52, 76, 77]. One approach is trimming, which generally refers to the process of manipulating an analog circuit's configuration in programmable discrete steps. One possible realization for temperature sensors is by adding a reconfigurable bias current source to the differential circuit to add an overcompensation to counter-balance any device mismatch. Addition of a reconfigurable resistor to fine-tune the output voltages of the sensor circuit is another approach. A relatively lower cost correction method is to adjust the amplification factor α that is used to derive V_{REF} from $V_{BE.}$ Time domain DTSs using delay lines also suffer from non-idealities.

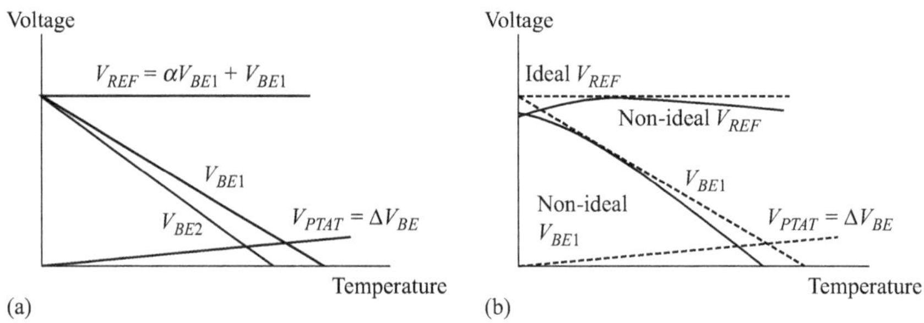

Figure 2.13 Contrasting the ideal and non-ideal diode behavior and its impact on sensor operation: (a) behavior of the forward-biased diode voltages and the bandgap reference voltage assuming ideal linear temperature dependency of diode voltages. (b) Curvatures introduced to the diode voltages and bandgap reference voltage due to non-linear relationship with temperature

The temperature dependence of the delay lines deviate from the ideal linear assumption resulting in curvature in their behavior. This error component can also be compensated in a similar fashion by adjusting the curvature of the corresponding reference delay line. This involves modification of the cells used in the reference delay line by sizing to introduce a counteracting curvature against the non-linearity experienced in the temperature-dependent delay line [52].

Overall, the correction methods either introduce non-negligible area and power overhead to the individual sensor design or they are simpler yet require additional effort during calibration.

2.5.2 Robustness against variations

While there are solutions with some associated area overhead to minimize the impact of non-idealities in analog BJT-based sensors, they still face a challenge due to the technology trends in future ICs. BJT structures do not scale robustly to keep up with the shrinking nodes of the CMOS processes and their proportionally dropping supply voltage levels used by the rest of the CMOS logic. For extremely small footprint and low power sensor designs targeting CMOS process nodes at 45 nm and below, these sensors will become difficult to interface with supply voltages dropping below 1.2 V, which is ideal for these analog sensors. One early proposal to mitigate this issue was to introduce a combination of a charge pump, voltage-controlled oscillator, and a voltage regulator to boost the analog thermal sensor's supply with respect to the supply level of CMOS logic [12]. This observation in fact has led to increased interest in designing CMOS sensor circuits.

As for bimetallic thermocouples, they are process invariant, since the thermovoltage is an intrinsic property of the metals. Only the single absolute reference point sensor will be built using a conventional method and hence, it will face challenges of that particular design style. On the other hand, this sensor system requires introduction of two new types of metals into the CMOS process and design of additional vias. Copper could be used as one of the thermocouple strips, which is already part of the CMOS process. In that case additional complexity would be less.

Yield, in conjunction with process variation, is a challenge for DTSs as well. In early processors there were only a few (2–3) sensors, which could be designed with less emphasis on area efficiency and circuits could be made robust to variation with the given area headroom. Also, yield would be minimally impacted by the contribution of a component that is represented at low numbers in the bigger scheme of things. Over the newer generations of processors and the growing importance of thermal monitoring particularly in multicore processor chips, trends have changed. High-end multicore processor chips contain a much higher number of thermal sensors [17] and their yield can now become noticeable. Furthermore, at advanced process nodes where most DTSs are now built, random variations in device parameters such as the threshold voltage V_{th} and carrier mobility μ become increasingly more pronounced. Building thermal sensors compliant with the yield expectations of the rest of the digital system is one prominent design challenge.

Figure 2.14 illustrates a methodology to carry out yield analysis of a thermal diode-based analog temperature sensor [78]. In this example, a total number of 1000 chips with on-die temperature sensors were examined subject to process and temperature variations. The sensor behavior is simulated using a circuit-level

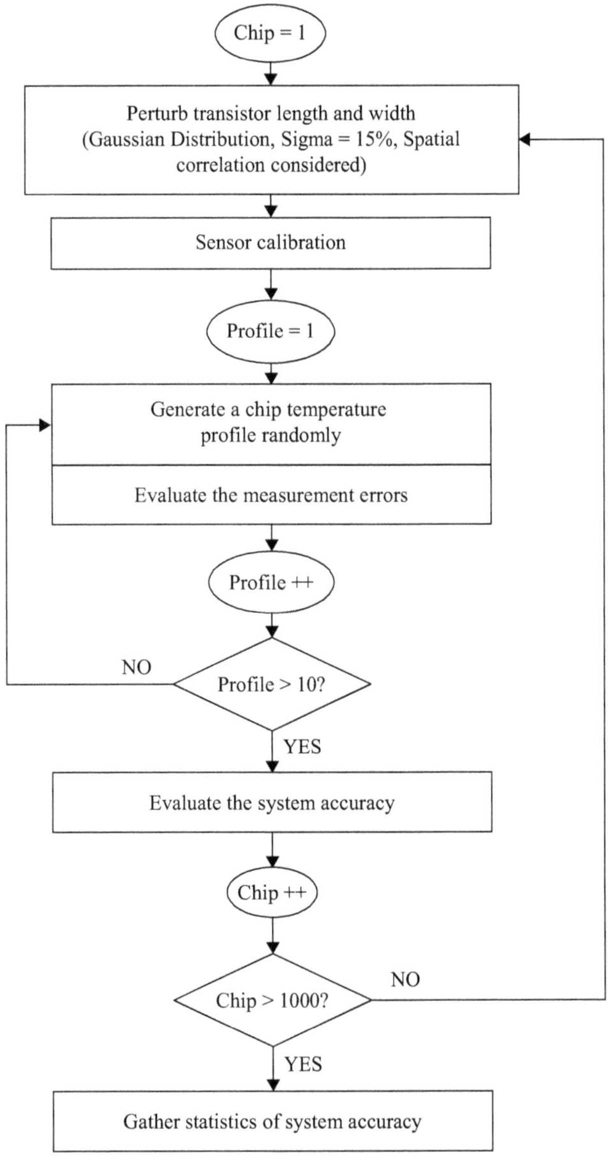

Figure 2.14 A flow for Monte Carlo analysis to assess the yield of a given sensor design under process variations

simulation tool (e.g., HSPICE) with appropriate technology models (IBM 10SF 65 nm technology models were used in this case study). In the simulation, the dimensions of each chip are fixed to a typical size of a representative design and the area has been divided into four quadrants and with one thermal diode allocated at the center of each quadrant. The thermal diode used in this particular study is a MOSFET operating in its linear region. The Monte Carlo simulation would then consist of three steps:

1. **Parameter perturbation:** In the first step, the length and width of the transistors are perturbed, where spatial correlation can also be accounted for. This perturbation is usually assumed to be in the shape of a Gaussian distribution. The standard deviation of the distribution can be set to any expected level based on the knowledge of the process. In this example it was set to 15%.
2. **Sensor calibration:** In the second step, the thermal diodes are calibrated.
3. **Thermal monitoring accuracy evaluation:** In the third step, the accuracy of the thermal monitoring system is evaluated using randomly generated chip temperature profiles. In this example, the lower bound and upper bound of the temperature have been set to 35°C and 115°C, respectively. For each temperature profile, the temperature in the proximity of each diode was estimated based on its forward bias voltage obtained from the circuit simulation of the sensor followed by the calibration. The measurement error of each diode is defined as the absolute value of the difference between the projected temperature and the actual temperature in the proximity of the diode. Further, we define the system accuracy of the monitoring system implemented on a chip as the maximum measurement error among all on-chip diodes across all temperature profiles. The average/best/worst accuracy and the standard deviation of the accuracy can then be calculated for a batch of chips. In this case study, the average accuracy is defined as the average of the accuracy across 1000 chips. The best/worst accuracy and the standard deviation of the accuracy are defined similarly.

Figures 2.15 and 2.16 depict two possible process spreads that could be observed in a carefully optimized sensor network, which has been subject to two different calibration methods. The sensors analyzed in Figure 2.16 have been subjected to a more thorough calibration (we will discuss the calibration and its impact in the following section).

We can observe that depending on the design of the sensors, the impact of process variation can manifest itself at varying degrees of severity. In this case, this particular optimized design using a weaker calibration scheme achieves an average accuracy of around 1°C. The stronger calibration method helps to improve the average accuracy to less than 0.6°C.

Nonetheless, there are a few chips with poor system accuracy (larger than 3°C). This is because for these chips, the perturbed channel lengths of some transistors were too small (less than 25°nm). These transistors exhibit high order of non-linearity affecting the accuracy of the monitoring system.

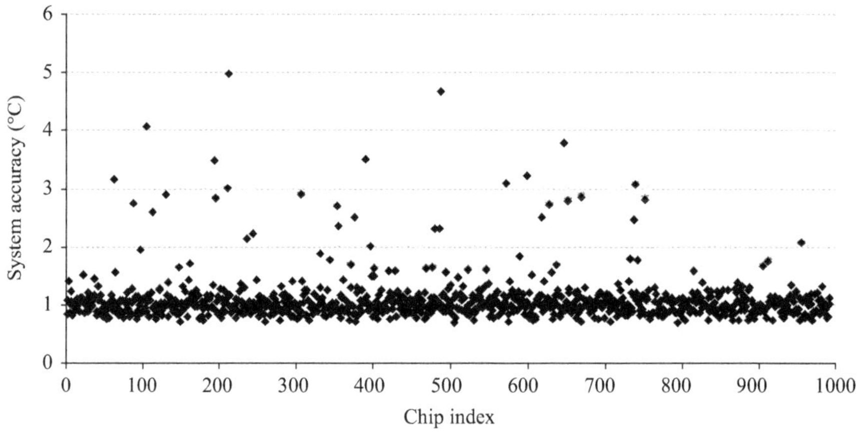

Figure 2.15 Monte Carlo simulation depicting the impact of process variation on analog thermal sensors integrated into a batch of chips. The size and power density of the underlying chip has been modeled after a high-performance processor

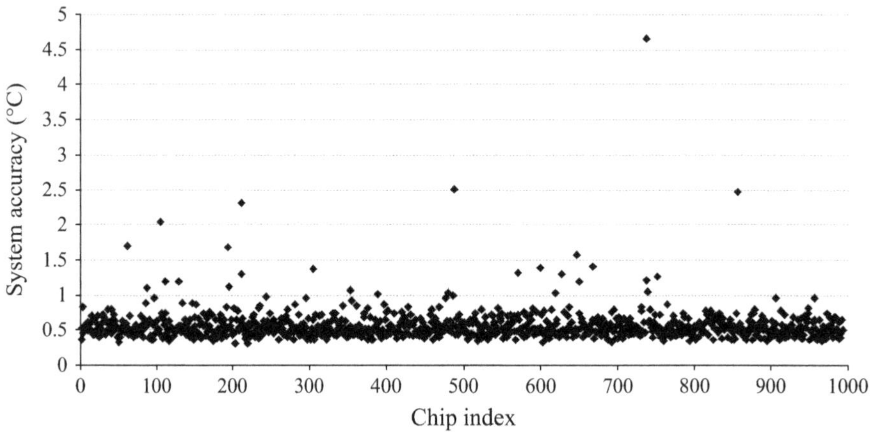

Figure 2.16 Monte Carlo simulation depicting the impact of process variation on analog thermal sensor integrated into a batch of chip. All simulation parameters are kept same as the one depicted in Figure 2.15, except for the calibration step. The calibration for this simulation utilized a more accurate (and costly) method

2.5.3 Calibration

Another major cost item for production of on-chip temperature sensors is the calibration process. Calibration is used to remove systematic bias in the measurement devices from the error components. For temperature sensors the goal is to adjust the raw response of the sensor so that the resulting actual temperature reading matches the true values at specific points of the Celsius scale with minimal offset. In some generations of processor chips, this cost has been viewed as so significant to the point that some products simply deployed uncalibrated sensors. Such monitoring systems relied on the assumption that the average case operating temperature range is not expected to threaten chip reliability. Furthermore, in these chips it is assumed that occasional exceeding of a temperature threshold leading to a delayed triggering of a thermal intervention mechanism is not likely to jeopardize the TDP rating of the chip. Nevertheless, this cannot be applied generally to all thermal sensors and chips they are in charge of monitoring and protecting. Especially, with the increasing contribution of cooling cost to the total energy cost and the trend to maximize power budgets through aggressive execution paradigms such as boosting, systems tend to operate much near the edge of thermal emergencies for more prolonged durations of time. As a result, fine granularity in spatial resolution as well as high accuracy is increasingly expected from on-chip temperature sensors. As a result calibration becomes highly relevant. Calibration techniques for temperature sensors mainly fall into two categories: single point and multi-point calibration.

The calibration process for temperature sensors consists of introducing an auxiliary means of verifying the true temperature of the system under observation with an external sensor and comparing the raw output of the sensor under test against this reference sensor. For different types of sensors, the procedure has slight variations.

During calibration, chips are placed in a temperature-controlled environment for measurements to be taken at known temperature points, where the chip can be maintained for a sufficiently long amount of time. At this known temperature the response from the sensors is collected. For analog sensors the voltage response from each of the sensors is measured. The temperature coefficient of each diode junction's voltage response is derived and programmed into fuses or special registers. Trimming can be applied to the sensor via a programmable resistance in series with the diode (BJT/MOSFET junction) to partially compensate for current source and temperature variation. Another important parameter to calibrate is the front-end ADC of the sensor device. One option used by commercial products [2] is to supply a high precision reference voltage from an off-chip source attached to the package during calibration. This voltage is carried over to the input of the ADCs through on-chip programmable resistor ladders. The converters cycle through a wide range of voltage values in small increments supplied from this source and the corresponding ADC output is tabulated against the corresponding input voltage. During actual operation the ADC outputs are extrapolated using the information stored in this table to derive an effective ADC output and map it to the ultimate temperature reading. Calibration for DTSs is performed during manufacturing tests using a similar method. For instance calibration of sensors that convert temperature to oscillator tracks the counts at known temperatures.

Calibration of resistive sensors involves measurement of the resistance at known temperatures and the derivation of a lookup table (LUT) or equation to relate the resistance of the device to temperature. The thermistors used in the IBM

POWER6 processor undergo such a procedure and the resulting linear fitting of the relationship is recorded as part of the processor vital product data and stored in an external electrically erasable programmable read-only memory (EEPROM) associated with each processor chip.

Fitting of resistance to temperature relationships for thermal diodes or resistive sensors should ideally be performed using a large number of data points. However, this procedure can become lengthy to apply to each single chip. Especially with the increasing number of on-chip sensors it is becoming cost prohibitive to perform individual calibration. In order to constrain this overhead, majority of temperature sensor calibration techniques use between one to three data points per sensors for linear fitting. A more sophisticated derivation could also use a piecewise linear fitting if the overhead is acceptable or the accuracy is needed for the specific application of the sensor.

Figure 2.17 illustrates the concepts of three-point and two-point calibration. In the three-point calibration process, the chip is placed under uniform temperature levels T_1, T_2, and T_3. The linear fitting can be either piecewise linear as shown in Figure 2.17(a) or linear as shown in Figure 2.17(b). A piecewise linear three-point calibration would be expected to best fit any non-linearities to provide highest accuracy. Linear fitting through a single calibration equation (shown in (2.7)) would be a simpler approximation.

$$T = slope \times Temp_{Coefficient} + Intercept \tag{2.7}$$

In the two-point calibration process, the chip is placed at two known temperatures T_1 and T_2. The forward bias voltages of each thermal diode at these two temperature levels are measured. Linear fittings for the voltage–temperature curve of each diode are determined based on these two measurement points. For two-point calibration, by taking measurements at two representative yet sufficiently diverse points in the operating range two data points can be obtained. Using these data points the expected linear behavior of the sensor is extrapolated and other points in the scale are inferred from a relationship such as the one depicted in (2.7). A study on industrial-grade sensors suggests using a pool of at least 30 sensors with three-point measurements [79] to create a sufficiently large population. The

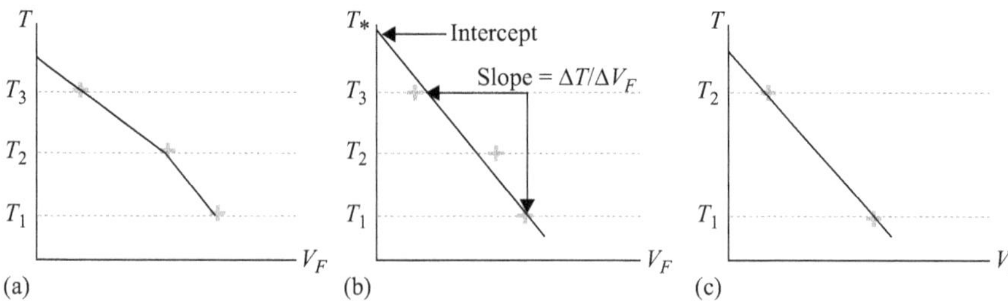

Figure 2.17 Illustration of sensor calibration: (a) three-point measurement with piecewise linear fitting, (b) three-point measurement with linear fitting, and (c) two-point measurement with linear fitting

regression applied to each individual sensor's three-point measurement yields one set of slope and intercept value. These are then averaged to obtain one representative linear equation.

One-point calibration has also been considered for certain sensor designs to minimize the testing overhead [72, 80–82]. It is argued that for certain sensor designs the sensor's response can be separated into two independent function, one involving a temperature-dependent parameter and the other involving a process variation-dependent parameter. If it can be demonstrated as such that the temperature-dependent coefficient of the response is to a large degree immune to process variations, then one-point calibration can be viable. Generally, calibration with fewer measurement points can more often yield acceptable accuracy for sensors with voltage/current outputs, while time domain sensors might not achieve the same level of accuracy with one-point or two-point calibration.

2.6 Improving temperature monitoring accuracy with systematic placement

Accuracy is crucial for thermal monitoring. Overestimation of temperature impacts performance negatively due to unnecessary triggering of thermal control mechanisms, for example, dynamic voltage and frequency scaling. On the other hand, the mean time to failure decreases exponentially with an increase in temperature. The quality of other circuit parameters also suffers with rising temperature as we have discussed earlier in section 2.1. Therefore, underestimation of the die temperature is also not desired since the chips will continue to operate at a higher temperature than their rated operating condition, hence greatly reducing the reliability.

An ideal method to allocate temperature sensors is to find the hottest regions over a set of well-defined applications and place one sensor in each location. Early attempts to create a dynamic monitoring for microprocessors employed a similar rationale. For example, for the Intel Pentium 4 processor, one sensor is placed near the rapid integer arithmetic logic unit (ALU), which has been determined to undergo the most severe thermal stress. Dual core 90 nm Intel Itanium processors utilized four on-die sensors placed near the integer and floating point units of each core [2]. On the other hand, such coarse predictions can be largely inaccurate. The thermal behavior of high power density chips such as microprocessors is affected by various factors. For example, localized heating on a processor is application dependent. In addition, process variations impact the total power consumption (by directly influencing the leakage component) and hence, varying thermal profiles across a batch of chips. Power management techniques, such as local clock gating and task migration, further create a disparity in power densities among different regions on a chip. Another important factor is the physical distance between a sensor and the location of thermal event. There is a maximum feasible temperature gradient between a hotspot of interest and a sensor placed a fixed distance from it. If the hotspot is beyond that range of reach of the sensor, the measurement can no longer be representative of that hotspot's thermal state. A study on the correlation between a sensor's distance from a temperature hotspot and its actual measurement established an exponential relationship as shown in (2.8) [83]. If the physical location of a sensor and a hotspot are separated by distance r, then the temperature of the hotspot perceived at the sensor's location can be expressed as:

$$T_{sensed}(r) = T_{hotspot}\left(1 - e^{-\frac{2r}{K}}\right) \tag{2.8}$$

In other words, the sensor will underestimate the temperature of the locations away from it, assuming that the distant hotspot is the hottest point on the chip and hence, it is of interest for tracking. The constant coefficient K is derived as a function of the effective thermal resistivity of the layers comprising the chip (silicon, heatspreader, heatsink, etc.) and their respective thicknesses. In presence of multiple heat sources with their respective hotspots, a superposition of the individual contributions to a thermal gradient between the sensor and each hotspot can be envisioned. However, the approximation will not be as accurate.

Considering the challenges in accurate tracking of hotspots, several studies focused on the nature of their distribution. It is intriguing to observe that several studies aiming to identify the hottest regions of microprocessors reached diverse conclusions. In earlier processor families, the few temperature sensors were placed near the integer and floating point execution units considering them as the likeliest candidates to cause a hotspot. For the Alpha 21364 architecture, the register file appears to be the hottest component consistently across a large set of SPEC CPU2000 [84] benchmarks [85]. Another experiment with the same basic architecture (with a slightly different configuration, mainly using different configurations of the memory hierarchy, particularly the level 2 cache blocks), benchmark suite, and thermal simulator revealed yet a different conclusion [86]. The issue queue (IntQ) generated the hottest points in most cases. Considering the wide variety of sensor allocations in commercial products and previously reported results, the conclusion is that the temperature behavior is architecture and workload dependent. This motivated the need for a systematic approach to the sensor allocation and placement problem for a given architectural configuration, workload characteristics, and other relevant design and system-level parameters.

One naive option to increase the accuracy is to place a very large number of sensors on the die. However, this incurs significant area, power, and testing cost due to several factors discussed earlier in this chapter. Main trade-off parameters between the two distinct types of sensors (analog vs. digital) is that digital sensors usually require more hardware overhead, while analog sensors generally have poorer compatibility with scaling due to their inability to keep up with lowering supply voltage levels and the process variations that exacerbate at deeply scaled process nodes. Compensation for these factors, in turn, makes the designs costlier to produce and test. Furthermore, sensors require an interface and a communication medium, such as inter-integrated circuit (I^2C), SPI, system management bus (SMBus), or a dedicated network on chip (NoC), which provides communication with embedded microcontrollers, off-chip controllers, and other digital systems. Allocating arbitrarily large number of such sensors not only creates a significant area overhead, but routing the data from the sensors to a central processing unit will also pose a challenge.

These observations led to the development of systematic methods for sensor allocation and placement. One of the first proposals defined the ultimate goal of an effective systematic technique as finding an allocation of thermal sensors and their physical locations to maximize the accuracy of the readings obtained from the placement while bounding the associated overheads [86].

2.6.1 Enhancing uniform grids with interpolation

One straightforward method to create a workload-independent sensor infrastructure is to divide the chip into equally sized grids and place a sensor at each grid point. Then, all sensors work in parallel, and the maximum temperature measured among the sensors will be used as the estimation of the core temperature. If the grid size is equal to the effective sensing radius of the thermal sensor, we can achieve 100% accuracy because each grid will have one dedicated sensor on it. However, it is usually not practical to implement such a fine-grain sensor grid. Every sensor incurs additional cost, which becomes non-negligible as the number of sensor increases. Furthermore, the placement of sensors into a highly optimized processor layout with a very limited white space will become complicated as the number of sensors increases.

Therefore, the static sensor grid size should be bounded, and at the same time, a metric is needed to evaluate the expected accuracy of such a static grid. The following key aspects are influential in the behavior of the static sensor grid formation. First, the accuracy obtained from a static grid placement is related to the number of sensors used; however, it is not a linear relationship.

Figure 2.18 presents the trend of the temperature tracking accuracy in relation to the number of sensors allocated in a regular grid on a single processor core

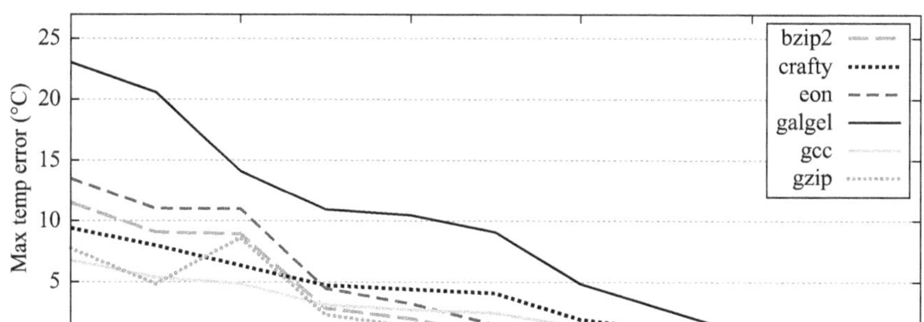

Figure 2.18 The relationship between the number of sensors allocated on a processor core and the worst-case temperature monitoring error. The x-axis is in log scale. Six different applications from the SPEC 2000 suite are depicted

executing representative computing benchmarks from the SPEC200 suite. We observe that when the number of sensor grids increases above 25 (note that the x-axis is in a log scale), the errors of sensor readings do not change any further. It is important to note that even if the error in tracking the maximum temperature decreases to a substantially low plateau for some applications rather quickly, for others, the error is still significant. Consider the galgel benchmark. Even with 64 sensors placed on a 8×8 grid on the core, the errors can be as high as 5°C. Considering that these readings will be used for the DTM and the thermal-emergency management, such incorrect readings can either invoke unnecessary performance throttling or put the processor in danger. The conclusion is that the size of the grid improves the effectiveness of the sensor infrastructure in many cases; however, in others, the hotspots may simply be located such that even a sizable grid of sensors will be incapable of capturing the locations of significant thermal events.

The inconsistent accuracy and potentially large errors in the basic uniform grid are caused by the fact that it is not trivial to optimize the distances of the sensor points to the hotspots by such a static placement. Therefore, in the presence of a uniform grid, a corrective measure is needed to further refine the readings obtain from the static uniform placement. The basic idea is that several sensor readings can be used to interpolate the thermal behavior of locations farther away from the actual physical sensors. Therefore, instead of using the sensor reading directly as an indicator of the hotspot temperature, a more accurate value of the highest temperature on the chip can be calculated by the thermal monitoring system by interpolating the sensor readings in the neighborhood of that hotspot.

An interpolation scheme was proposed to estimate hotspot temperatures using grid sensor readings for a multicore CMP [87]. For each sensor, its reading and those of its neighbors are used to estimate the position and temperature of the hottest spot within its neighborhood. The neighborhood of a sensor S_i is defined as

$$N(S_i) = \left\{(x, y) | \frac{x_i - r_s}{2} \le x \le \frac{x_i + r_s}{2}, \frac{y_i - r_s}{2} \le x \le \frac{y_i + r_s}{2}\right\}, \tag{2.9}$$

where (x_i, y_i) is the x-y coordinates of a given temperature sensor, S_i, and r_s is the distance between two neighboring sensors. This neighborhood represents a square region of with one side measuring r_s with sensor S_i at its center. Considering the sensors shown in Figure 2.18, the neighborhood of S_4 is the region inside the dashed square with sides extending midway between S_4 and its immediate neighbors, S_1, S_3, S_5, and S_7. Once the maximum temperature estimation within the neighborhood of each sensor is derived, the largest of the values reported among all sensors can be used as the peak temperature.

For the neighborhood depicted in Figure 2.19, let us consider the thermal profile along the x direction first. If the reading of S_5 is larger than those of S_3 and S_4, the hottest spot within $N(S_4)$ should be close to the mid-point between S_4 and S_5. Particularly, if the gradient of the thermal curve does not change between S_4 and S_5, the x coordinate of that hottest spot should be $x_4 + r_s/2$. Likewise, if the reading of S_3 is larger than those of S_4 and S_5, the x coordinate of the hottest spot within $N(S_4)$

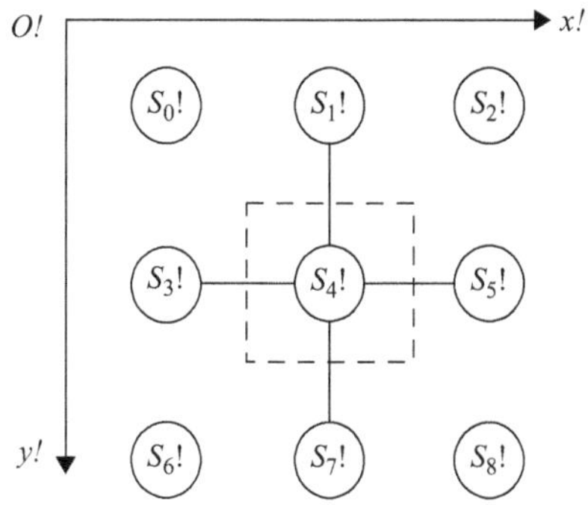

Figure 2.19 Illustration of the definition of a sensor's neighborhood for collective interpolation

can be estimated as $x_4 - r_s/2$. If the reading of S_4 is larger than those of S_3 and S_5, both of the gradients between (S_3–S_4) and (S_4–S_5) can be used to perform the estimation. It is assumed that the temperature gradient is constant within the segment between the hottest spot and the nearest sensor, that is, in that segment the temperature can be approximated as a linear function of x and y coordinates. Assuming the actual temperature readings of S_3, S_4, and S_5 are T_3, T_4, and T_5, respectively, the x coordinate of the hottest spot within $N(S_4)$ should be $x_4 + \frac{1}{2}\frac{\Delta T_3 - \Delta T_5}{\Delta T_3 + \Delta T_5} r_S$, where $\Delta T_3 = T_4 - T_3$, $\Delta T_5 = T_4 - T_5$.

Similar arguments apply to the y direction. In general, the coordinates of the hottest spot within $N(S_4)$ can be estimated using (2.10) and (2.11). Then, the estimated coordinates of the hottest spot can be substituted into the linear function representing temperature within the limited segment between the hottest spot and the nearest sensor location. This substitution results in (2.12) and (2.13), which are used to obtain the value of the estimated maximum temperature.

$$\Delta x = \begin{cases} \frac{1}{2}\frac{\Delta T_3 - \Delta T_5}{\Delta T_3 + \Delta T_5} r_S \ldots\ldots\ldots T_4 \geq \max(T_5, T_3) \\ -\frac{1}{2} r_S \ldots\ldots\ldots\ldots\ldots\ldots\ldots\ldots T_3 \geq \max(T_4, T_5) \\ +\frac{1}{2} r_S \ldots\ldots\ldots\ldots\ldots\ldots\ldots\ldots T_5 \geq \max(T_3, T_4) \end{cases} \tag{2.10}$$

$$\Delta y = \begin{cases} \frac{1}{2}\frac{\Delta T_1 - \Delta T_7}{\Delta T_1 + \Delta T_7} r_S \ldots\ldots\ldots T_4 \geq \max(T_7, T_1) \\ -\frac{1}{2} r_S \ldots\ldots\ldots\ldots\ldots\ldots\ldots\ldots T_1 \geq \max(T_4, T_7) \\ +\frac{1}{2} r_S \ldots\ldots\ldots\ldots\ldots\ldots\ldots\ldots T_7 \geq \max(T_1, T_4) \end{cases}$$

$$(x_{est}, y_{est}) = (x_4 + \Delta x, y_4 + \Delta y) \tag{2.11}$$

$$\Delta T_x = \begin{cases} \frac{|\Delta T_5 - \Delta T_3|}{\Delta T_5 + \Delta T_3} \max(\Delta T_5, \Delta T_3) \ldots\ldots\ldots\ldots T_4 > \max(T_5, T_3) \\ \frac{\max(T_5, T_3)}{2} - T_4 \ldots\ldots\ldots\ldots\ldots\ldots\ldots\ldots T_4 \leq \max(T_5, T_3) \end{cases} \quad (2.12)$$

$$\Delta T_y = \begin{cases} \frac{|\Delta T_7 - \Delta T_1|}{\Delta T_7 + \Delta T_1} \max(\Delta T_7, \Delta T_1) \ldots\ldots\ldots\ldots T_4 > \max(T_7, T_1) \\ \frac{\max(T_7, T_1)}{2} - T_4 \ldots\ldots\ldots\ldots\ldots\ldots\ldots\ldots T_4 \leq \max(T_7, T_1) \end{cases}$$

$$T_{N(S_4)}^{max} = T_4 + \Delta T_x + \Delta T_y \quad (2.13)$$

Equations (2.10)–(2.13) are based on the assumption that the gradient of the temperature curve remains constant between S_4 and the estimated hottest spot. However, on a real chip, this may not be the case. Around certain temperature peaks, the temperature gradient can change significantly. Therefore, the peak temperature could be largely overestimated if these equations are applied directly. Equations (2.10)–(2.13) can be modified to take this into account resulting in the following equations:

$$\Delta T_x = \begin{cases} \kappa \frac{|\Delta T_5 - \Delta T_3|}{\Delta T_5 + \Delta T_3} \max(\Delta T_5, \Delta T_3) \ldots\ldots\ldots\ldots T_4 > \max(T_5, T_3) \\ \beta(\frac{\max(T_5, T_3)}{2} - T_4) \ldots\ldots\ldots\ldots\ldots\ldots\ldots\ldots T_4 \leq \max(T_5, T_3) \end{cases} \quad (2.14)$$

$$\Delta T_y = \begin{cases} \kappa \frac{|\Delta T_7 - \Delta T_1|}{\Delta T_7 + \Delta T_1} \max(\Delta T_7, \Delta T_1) \ldots\ldots\ldots\ldots T_4 > \max(T_7, T_1) \\ \beta(\frac{\max(T_7, T_1)}{2} - T_4) \ldots\ldots\ldots\ldots\ldots\ldots\ldots\ldots T_4 \leq \max(T_7, T_1) \end{cases} \quad (2.15)$$

The physical meaning of the coefficientκ used in (2.14) and (2.15) is illustrated in Figure 2.20(a). When $T_5 > T_3$, κ is the average thermal gradient between x_4 and x_{est} divided by the average gradient between x_3 and x_4 (if $T_3 > T_5$, then, κ should be the average gradient of the thermal curve between x_4 and x_{est} divided by the average gradient between x_5 and x_4). The physical explanation of the coefficient β is similar.

If it can be assumed that the chip is thermally isotropic, then the same κ and β values for both x and y directions can be used. Generally speaking, calculating the value of κ and β is nontrivial since it is hard to find an exact equation to fit the temperature–distance curve. However, it can be seen from Figure 2.20 that κ should be close to zero since the gradient of the temperature curve approaches to zero quickly near the maximal point. On the other hand, β should be close to one, as the gradient does not change much between x_4 and x_5. Although κ and β are complicated functions of sensor location (x_i, y_i) and sensor's distance r_s to expected points

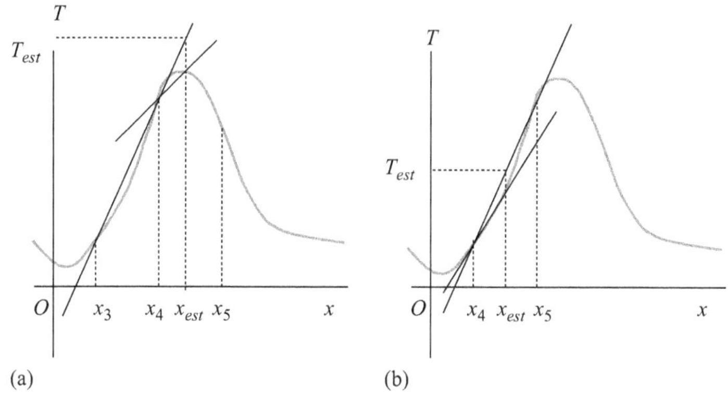

Figure 2.20 Physical depiction of (a) κ and (b) β

of thermal events, and also are architecture dependent, empirically it was found that setting κ to $0.83/\sqrt{N_{sensor}}$ and β to 0.93 has given the smallest estimation error, where N_{sensor} is the number of sensors allocated per core.

Based on this interpolation scheme, a sensor management method called interpolation-based dynamic sensor selection was developed [87]. This method coordinates a large number of on-chip sensors with minimal operation overhead. A large number of grid sensors are embedded into each core, however, the thermal monitoring unit receives readings from only a small subset of them at a given instant. The idea is to create a hierarchy of sensors and first activate coarse-grain sensors and use their readings to estimate the hotspot location of the core using the interpolation technique. At this time all other sensors and their related support circuitry can be put to an idle/sleep state. Then, the grid sensor closest to the estimated hotspot location is activated. To further increase the accuracy, eight sensors surrounding this sensor are also activated since the location estimation may not be precise. The maximum among the readings of these nine sensors are used as the estimated hotspot temperature of the core. By directly reading the grid sensor instead of calculating temperature indirectly helps to adapt better to different architectures.

The methodology is depicted in Figure 2.21. In this example, let us assume that the total number of the grid sensors is 64 (only a subset is shown here). The chip is divided into 16 equal-sized quadrants, which use grid sensors S_0–S_{15} residing at the center of the quadrants, serving as the coarse-grain sensors. These coarse-grain sensors are activated in the first stage. Their readings will be fed into (2.10) and (2.11) to estimate the hotspot location. In Figure 2.21(a), the estimated hotspot location is marked with a star. In the second stage, the grid sensor that is closest to the estimated location and eight sensors surrounding it are activated, as shown in Figure 2.21(b). The maximum reading from these nine activated sensors will be used directly as the hotspot's temperature. In this example, the total number of activated sensors is only 25 (16 coarse-grain sensors plus nine grid sensors that are activated later).

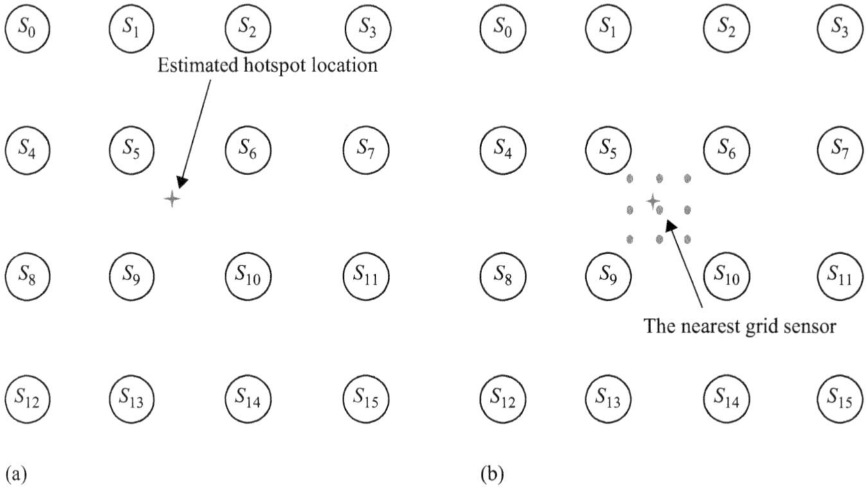

Figure 2.21 Steps for the interpolation-based dynamic selection method: (a) 16 coarse-grain sensors activated, hotspot location estimated based on the coarse-grain sensor reading and (b) the grid sensor closest to the estimated hotspot and 8 surrounding sensors are activated

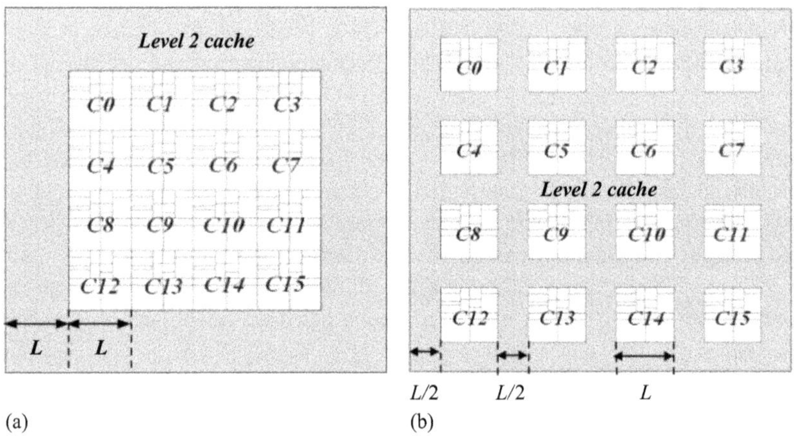

Figure 2.22 Floorplans of the CMP chip for (a) the dense architecture with L2 cache surrounding the array of cores and (b) the sparse architecture with L2 cache distributed between the cores

The interpolation method has been tested on two CMP architectures, which are referred to as the dense and the sparse architectures due to the placement of cores in each. Figure 2.22 depicts these two alternatives. The dense architecture's floorplan packs all cores tightly with the entire L2 cache surrounding the cores, while the sparse architecture spreads the cores allocating the space between them for the L2 cache. Naturally, the dense architecture is expected to undergo higher thermal stress.

It is relevant to mention the evaluation setup for this study. A grid size of 288 × 288 for the entire chip has been used for the dense architecture, and a 312 × 312 grid was used for the sparse architecture, both corresponding to a sensor grid size of 48 × 48 per core. With a steady-state thermal analysis the initial temperature for each chip is determined for the subsequent transient analysis. This represents the warm-up period. The ambient temperature is set to 40°C. Subsequent simulations are divided into intervals corresponding to 0.05 s of execution. For a 2.5 GHz clock frequency this corresponds to a sampling period of every 125 million cycles, which constitutes a simulation interval. The entire thermal simulation lasts for 15 simulation intervals (1.875 billion cycles). The initial temperature at the start of each interval is taken from the final temperature of the previous iteration. When the maximum temperature of any core is above the emergency temperature (which is set to 82°C in this experiment), the chip has been throttled using a model for dynamic voltage and frequency scaling. The dynamic power output from the power simulator was scaled accordingly. Based on the scaled dynamic power, the temperature and the leakage power were calculated using an iterative method until the leakage power converges. An interesting observation of this study is that when using the interpolation method, for the same number of sensors per core, the monitoring error does not change much between the architectures. The interpolation method enjoys higher level of portability. Another conclusion drawn was that the dynamic selection scheme results in similar accuracy with a large reduction in number of sensors involved in data collection over a static uniform grid-based scheme. For instance, for both architectures, when dynamically activating only 18 out of 36 sensors, dynamic selection scheme yields almost the same accuracy and success rate in detection of thermal emergencies as a static 6 × 6 grid.

2.6.2 Non-uniform sensor allocation and placement

An alternative to building a static and uniform grid is to utilize profiling information on the target application domain to construct a custom, non-uniform sensor allocation and placement. One straightforward approach is detecting the potential hotspots through simulation and then placing a thermal sensor near each hotspot. This method has several problems in practice. First, the locations of hotspots are highly application dependent; hence, an optimal location for one application will not be the best solution for another. Second, the number of hotspots can be very large if we run a sufficiently large number of benchmarks to reveal all possible hottest locations on the chip across all target applications.

Therefore, it is necessary to derive an automated scheme to decide the optimized allocation of a set of sensors. The first systematic procedure proposed for this problem is carried out in two stages [86]. The first step is to generate a full thermal hotspot map across a wide range of applications. After that, the problem can be formulated as a clustering of the points of interest in the spatial domain. The number of clusters can be decided by the number of available sensors or it can be optimized through an iterative process of trying a range of sensor allocation budgets. The center of each cluster will indicate the physical location of a sensor. This sensor will monitor the points associated with that cluster. Hence, the temperature reading from that sensor is representative of its respective coverage area. The reading error of each sensor can be represented by the distance between the cluster center and the other points in the cluster. Therefore, minimizing the total reading error is formulated as a variant of the *k*-means clustering problem [88, 89].

2.6.2.1 Basics of *k*-means clustering

The basic *k*-means clustering technique can be defined as follows:

Given an integer k and a set of n data points $R = [\vec{a}_i/(x_{i1}, x_{i2}, \ldots, x_{im}), i = 1, 2, \ldots, n]$ each represented with a coordinate vector $\vec{a}_i$ in an m-dimensional space, determine k centers such that the mean-square distance from each data point to its nearest center is minimized, that is,

$$\text{Minimize} \sum_{i=1}^{n} |\vec{a}_i - \vec{c}_i|^2$$

where $\vec{c}_i$ represents the nearest cluster center to the data node $\vec{a}_i$.

The *k*-means clustering algorithm works by iteratively refining the position of the k cluster centers. Initially, the k cluster centers are randomly picked up from the n data points. Then, each data point finds out which center it is closest to. In this way, each cluster center will "own" a set of data points. The next step is to move the cluster center to the centroid of the points it owns. The iterations will continue until the optimal solution is reached.

The termination of the basic *k*-means algorithm is easy to prove. Each iteration will arrive at a new clustering configuration because the objective function value must be reduced. On the other hand, there are only a finite number of ways of partitioning the n data points into k groups. Therefore, the algorithm will terminate eventually. The algorithm can be modified to terminate sooner by either setting a hard limit on the number of iterations or on the incremental improvement observed in the objective function.

However, it is not possible to guarantee the optimality of the basic *k*-means algorithm [90]. Therefore, some guidance must be applied to the initial random distribution. One effective heuristic proposed to this end is to place jth initial center on the data points that are as far away as possible from the closest of centers 1 through $j - 1$ [90]. For the specific purpose of temperature sensor allocation, k sensors (k clusters) need to be created to monitor n hotspots. The cluster center corresponds to the sensor location. The data points in each cluster correspond to the hotspots that will be monitored by the sensor located at the cluster center.

2.6.2.2 Thermal-gradient-aware (TGA) *k*-means clustering

The basic *k*-means algorithm correlates the sensor error with the distance between the hotspot and the location of sensor. This implies a linear relationship between the temperature gradient around a hotspot and the temperature at that location. Therefore, if we have, for example, one point with the highest temperature of 370 K and another point with the highest possible temperature of 350 K, the basic *k*-means algorithm will place the sensor in the middle between these two hotspots. However, it is beneficial in terms of accuracy if we move the sensor toward the 370-K hotspot because the temperature gradient around a high-temperature location is larger than that at a low-temperature point. Therefore, a better sensor allocation should place an emphasis on minimizing the actual distance weighted by the temperature of the hotspots instead of using physical distances directly. A TGA *k*-means clustering and a sensor-allocation algorithm can overcome this challenge.

Considering the thermal characteristic *T*, that is, the temperature, of the hotspots, each such element can be regarded as distributed in a 3D space and described by a tuple (x, y, T). By using this representation, the gradient-aware sensor-allocation algorithm operates in two stages. In the first stage, the hotspots are grouped into clusters where elements in the same cluster exhibit both spatial and thermal correlation. In the second stage, the physical location within each such cluster is identified where a thermal sensor should be placed. The sensor placed at this location would provide the most reliable information regarding the thermal condition of any hotspot within a certain cluster. If the sensor is placed directly at the center of the cluster, it is called a 3D-placement scheme; otherwise, the location can be decided by a heuristic, as will be discussed later.

The clustering stage in a 3D space is similar to the basic *k*-means iterations in a 2D space, except that the 2D Euclidean distance needs to be extended to a 3D representation. A transformation shown in (2.10) is used to find the cluster center to which each hotspot belongs:

$$d(i,j) = (x_i - x_j)^2 + (y_i - y_j)^2 + (T_i - T_j)^2 \tag{2.10}$$

where (x_i, y_i) and (x_j, y_j) are the coordinates of the hotspot and the cluster center, respectively, and (T_i, T_j) denote the temperature levels of the hotspot and cluster center.

Once the hotspots are clustered into *k* groups, the physical locations of the temperature sensors will be determined. Instead of directly using the centroid of each cluster as the sensor location, this method proposes a different approach that takes the diversity of thermal gradients within a cluster into account. The basic idea behind this approach is to move the cluster centers or the sensors closer to the relatively higher temperature hotspots. This is equivalent to the sensor being attracted to the hotspots with high-temperature values with a larger force. The details are described in the algorithm shown in Figure 2.22.

The 3D Euclidean distance computation shown in (2.9) corresponds to Step (5) of the algorithm. If the temperature of a certain hotspot is larger than the average of the cluster, the sensor location will be pushed toward that hotspot by an attraction coefficient α. This is shown in Step (13). It has been determined empirically that an attraction coefficient value $\alpha = 0.1$ performs best. The cluster centers determined in Steps (3)–(14) are then iteratively refined such that the mean square distances of the hotspots from their respective cluster center are minimized over Steps (1)–(18). Note that computing the cluster center using this method moves the sensor location physically closer to the steeper thermal gradients. We should also point to the fact that although the temperature dimension of the cluster centers has been used in the distance definition, this dimension *T* is used only for modeling attraction toward points exhibiting high thermal stress. The temperature at the sensor's physical location determines its thermal reading, which has no physical relation with the temperature coordinate T_j of the cluster center.

Finally, it can be noted that the n-dimensional clustering concept allows this method to consider other parameters during the sensor placement. In addition to the thermal gradients, frequency of occurrence of a certain hotspot may be another important parameter to guide the placement. This parameter can be incorporated as an additional dimension defining each thermal event point in the cluster sets. The basic algorithmic approach outlined above can be embodied in a number of strategies as follows [91]:

Global sensor allocation: In this strategy, the global (i.e., chip-wide) hotspots are considered. Global hotspots are those locations where the absolute highest temperatures are observed for a given workload across the entire chip. The global hotspots generally emerge in the same functional block repeatedly over many applications for a given architecture configuration (although the exact location of the hotspot may be shifted inside the block across applications). However, there can be reasons for the global hotspots to move into different components of a chip. For instance, in a superscalar processor chip, multiple copies of the floating-point unit can be selectively activated. During the intervals where a single floating-point unit is active, it can contain the hotspot. In another interval where multiple copies of this unit are active, the load would be distributed evenly, and the power density would be low in this location. At that point, a different unit, such as the instruction queue, can be the origin of the hotspot. The global sensor allocation strategy places the sensors to capture such events as follows. First, an initial number of sensors are estimated, and a TGA sensor allocation is performed with that number. Then, the sensor allocation is modified iteratively until the resulting thermal monitoring accuracy falls within a given range. A good starting point is to select the number of sensors to be equal to the total number of major blocks and then to increase or decrease that number of sensors.

Local sensor allocation: In this case, the goal is to determine the allocation of the sensors for each individual processor component or block. Hotspots are redefined on a per component basis in this case. For a component, a hotspot is the location exhibiting the highest temperature observed within that block for an application. Effective local monitoring can be vital in various dynamic optimizations. For instance, activity-migration and thread assignment techniques can be assisted by local thermal monitoring mechanisms. Temperature information regarding local components can also be exploited by dynamic cache optimizations [92–94] to reduce the leakage power. The peak temperatures observed in memory blocks are never as high as some of the other blocks and they are unlikely to exceed safe temperature thresholds to trigger performance throttling. However, the leakage power of static random access memory (SRAM)-based memory blocks have a strong dependency on their thermal state and the increase in leakage at even relatively small gradients can become significant.

There can be different approaches for local sensor allocation:

Naive allocation: The most straightforward approach is to place a fixed number of sensors per processor block. There are different ways to place the sensors based on the geometry and alignment of the block. The main idea is to recursively bisect the block into smaller units until the number of units is equal to the number of desired sensors. For example, this will involve placing a single sensor at the geometric center of the processor block. For two sensors, the block is bisected along the longest edge, and a sensor is placed at the center of each bisected rectangle.

Single sensor at TGA-weighted cluster center (WCC): This technique involves placing a single sensor for each subcomponent. This is equivalent to applying the thermal-gradient aware allocation shown in Figure 2.23, without performing clustering within the component block. In this case, the entire hotspot map will form a single cluster, and the center of this cluster will be the sensor location for the single sensor. Hence, this method is termed as the TGA WCC. It is the same as the TGA k-means clustering approach but with k being equal to one.

TGA allocation: In this approach, multiple sensors are placed within each block. Such allocation is performed by the TGA k-means algorithm, where k is the number of sensors in each block, and the centers of the k clusters are the location for the k sensors. Increasing the number of sensors increases the monitoring accuracy.

Hybrid (HY) sensor allocation: In contrast to the previous approaches where all blocks have an equal number of sensors (either a single sensor for the WCC

```
Algorithm k–means clustering ( )

Input   : hot_{x,y,t} [N] : array of hotspots locations
          cluster_{x,y,t} [K] : array of sensor locations
Output  : member [N] : array of membership of hotspots

Initialize cluster [K] to be K points in hot [N]
WHILE δ/N > threshold
   δ ← 0
   FOR i=0 TO N-1
      FOR j=0 TO K-1
           d ←  |hot_{x,y,t}[i]-cluster_{x,y,t}[j]|
           IF d < d_min
              d ← d_min
              n ← j
       IF member[i] ≠ n
          δ ← δ +1
          member[i] ← n
      newsize[n] ←newsize[n]+1
      new_cluster_{x,y}[n] ← new_cluster_{x,y}[n]+hot_{x,y}[i]
      + α (new_cluste r_{x,y}[n]-hot_{x,y}[i])
      × (hot[i(t)]-new_cluster[n(t)]/newsize[n])
      new_cluster_t[n] ← new_cluster_t[n]+hot_t[i]
   FOR j=0 TO K-1
      cluster_{x,y,t}[j] ← new_cluster_{x,y,t}[j]/newsize[j]
      new_cluster_{x,y,t}[j] ← 0
      newsize[j] ← 0
```

Figure 2.23 Pseudocode of the TGA k-means clustering algorithm for sensor placement

strategy or a fixed number for the TGA strategy), the HY strategy allows the allocation of variable number of sensors in different blocks. The amount of sensors necessary for each subcomponent can be determined based on the error in monitoring observed in different blocks. At first, for each block, a single sensor is allocated by the WCC method, and the temperature monitoring error is determined. Then, depending on the degree of error observed in each individual block, the TGA strategy is applied repeatedly with an increasing number of sensors until no significant improvement in accuracy can be observed in that block. By customizing the number of sensors required for accurate temperature monitoring for each block, the total number of sensors is reduced.

Figure 2.24 illustrates an example of sensor placement that would result from the *k*-mean clustering-based method. Figure 2.25 depicts the performance of three variations to the clustering algorithm applied for global allocation: 2D (using the simple distance metric), 3D (incorporating temperature as a third dimension), and TGA (enhancing the algorithm with an attraction force towards steeper gradients). The underlying chip is modeled after an Alpha-like processor core and its power and thermal behavior was simulated with the cycle accurate architectural simulator Gem5 [95], Wattch [96], and Hotspot [97].

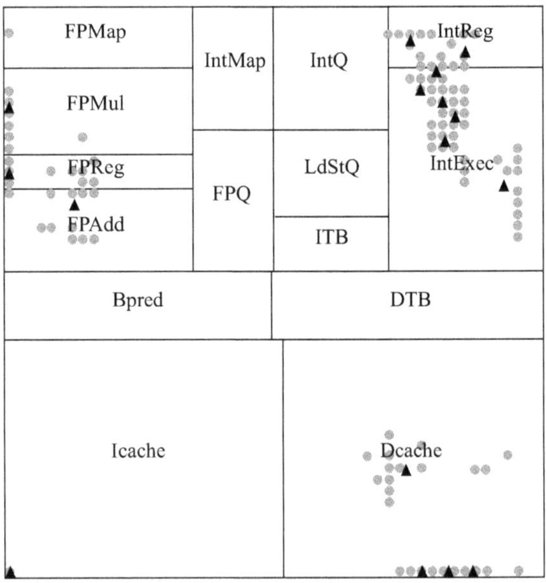

Figure 2.24 Sensor locations derived from a collection of candidate hotspots observed from thermal simulations of a set of workloads on a representative processor core. The orange dots represent all locations in the core where the peak temperature for a given run of a workload was observed. The dark triangles depict the sensor locations identified from TGA k-means clustering

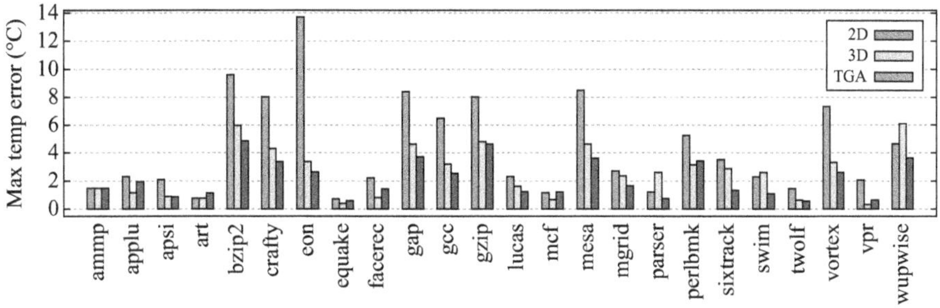

Figure 2.25 Maximum temperature error for the SPEC2000 benchmarks performing a non-uniform global sensor allocation using 2D allocation, 3D allocation, and TGA approaches

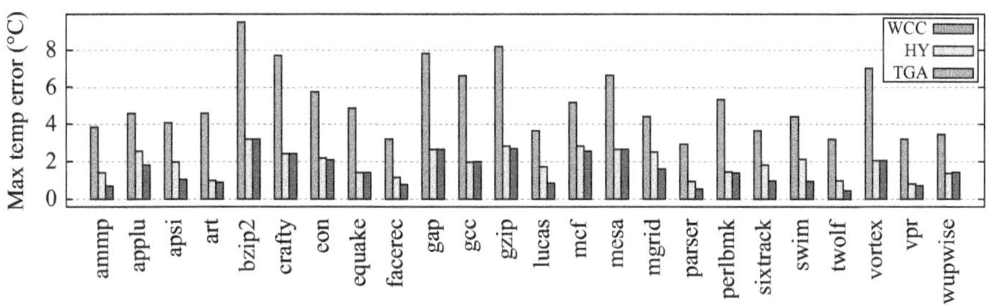

Figure 2.26 Maximum temperature error for the SPEC2000 benchmarks performing a non-uniform local sensor allocation using WCC approach, HY approach, and TGA applied at the locally at a subcomponent level

Figure 2.26 shows the comparison between three local allocation strategies: WCC, TGA, and the HY method. The local blocks in this case are the micro-architectural blocks of a processor, such as execution units, register files, various queue structures, private L1 data, and instruction caches.

2.6.3 Sensor allocation and placement for reconfigurable fabrics

Thermal concerns have a wide impact on computing platforms beyond micro-processors. Memory devices and reconfigurable devices can also experience high power densities and temperature has adverse effects on their performance and reliability as well. Optimization of temperature sensor placement on reconfigurable devices has been investigated with this motivation. For commercial-grade field programmable gate arrays (FPGAs) the maximum die temperature without performance degradation is reported as 80°C and the absolute safety limit is 125°C [98]. An average design mapped onto an FPGA with 90% device utilization could lead to a die temperature that is 50°C above ambient temperature. Considering the temperature of the case to be 40°C, the die temperature would be 90°C in still air and with a fan it would drop to 75°C [98]. Preventive measures for convective cooling (such as large heat sinks) or active cooling (such as fans) may not be feasible to deploy in embedded applications where FPGAs are often deployed due to size constraints. Therefore, the operating conditions can be expected to remain well above 75°C under high activity.

Thermal monitoring for FPGAs is relevant for two reasons: (a) FPGAs dissipate a significant amount of power and their operating temperature can exceed their rated critical die temperature in the absence of elaborate cooling mechanisms. The high power dissipation trend and consequent thermal stress has become more severe for technology nodes at 65 nm and below; (b) FPGAs are often used for rapid prototyping and emulation. Gathering thermal data from the FPGA device is important to characterize the design being emulated.

The programmability of FPGAs presents an opportunity to tailor temperature sensor locations after the particular application, which will utilize the device, is known. The temperature sensors can be generated using the native resources of the FPGA device, that is, configurable logic blocks (CLBs). At the same time, the exact utilization of programmable components and resulting physical factors such as power and temperature are not known a priori, since the same FPGA device can be programmed to perform various tasks. The CLBs can be occupied or unoccupied depending on the particular application. The power distribution and power density observed on the same FPGA device can be different for two different applications. As a result, locations of hotspots are application dependent. Moreover modern FPGAs contain embedded digital signal processing (DSP) blocks and microprocessor cores. These embedded cores can either remain unused or exhibit localized heating if they are utilized by the application. There can be multiple hotspots in the FPGA due to the uneven activity at different parts of the homogenous logic array and stripes of embedded cores.

Programmability can be exploited for embedding thermal sensors into an application immediately before mapping it onto the target device. Once the application to be mapped is determined, the candidate hotspot locations can be identified and sensors can be instantiated using unoccupied reconfigurable logic on the FPGA device. Ring oscillators-based DTSs have been demonstrated as effective means for thermal monitoring for FPGAs [60, 99].

Static placement of ring oscillator-based sensors requires significant amount of resources and incurs communication overhead to read back values from sensors. One technique to circumvent the resource overhead is to perform dynamic reconfiguration to insert sensors, take readings, and then vacate the resources occupied by the sensors for the use of the application [99]. For a large number of sensors distributed across a large FPGA device this will incur significant overhead, especially if thermal monitoring is used for taking preventive measures during the run-time of the application. Therefore, it is important to reduce the number of sensors and control their locations (while maintaining a given level of monitoring accuracy) in order to reduce the run-time reconfiguration and read back delay in such a scenario. Using less real estate to implement sensors may even allow static placement. This also simplifies the microcontroller and peripheral design used to control such statically placed sensors.

The first proposal for systematic placement of temperature sensors in FPGAs aimed to create a set of circular regions C where each circle $c_i \in C$ is centered around a profiled hotspot h_i and has a radius r_i. Then, the problem is formulated as identifying the minimum number of available CLB locations where a sensor can be instantiated, such that at least one sensor falls into each c_i region [100]. The radii of the circular regions are determined using the exponential model shown in (2.8), solved for r:

$$r_i = 0.5Kln\left(\frac{T(h_i)}{T(h_i) - \Delta T}\right) \quad (2.11)$$

Figure 2.27 demonstrates how this algorithm works on a logic array. The circles around hotspots represent their respective ranges. The shaded CLBs within the circles represent feasible sensor positions with unoccupied CLBs available that cover the respective hotspot. In this example sensor s_1 covers hotspots (h_1, h_2, h_3) and sensor s_2 covers (h_4, h_5).

Another method first considers placing a regular array of sensors within a fixed grid across the entire CLB array and this placement is referred to as the grid-based placement. This is equivalent to the logic array being partitioned into grids equal to the number of sensors and each sensor being placed at the center of the grid. This grid block is called the coverage area of the associated sensor. An illustration of grid-based placement is shown in Figure 2.28(a). The temperature measured by a sensor placed at the center of the coverage area shall be representative of the

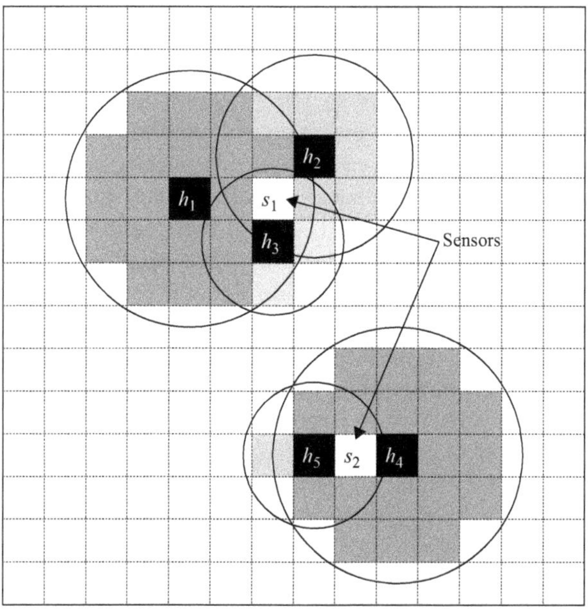

Figure 2.27 Demonstration of the covering of a set of hotspots by a smaller set of sensors utilizing the circular coverage region around the hotspots

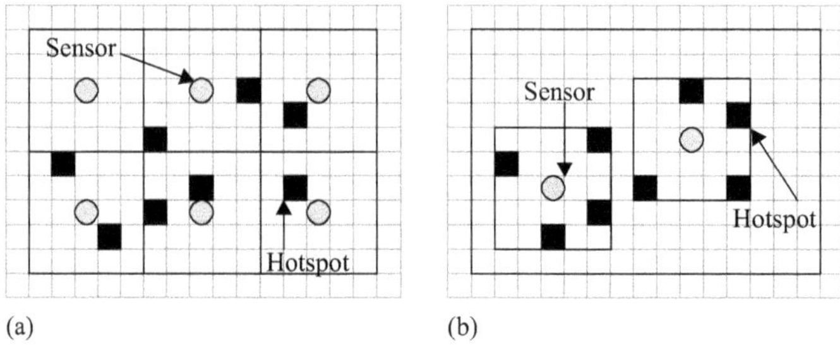

Figure 2.28 (a) Grid-based placement of sensors around the coverage perimeter of a. (b) Minimal sensor placement for a set of hotspots. The rectangular coverage area of each sensor is covering four hotspots

temperature of the entire coverage area. However, different points within the coverage area can be at different temperatures. The error in approximating the temperature of the coverage area by the temperature measurement at its center is tolerable if it is below a given error margin ΔT. The largest distance away from the sensor at which the approximation error is less than ΔT determines the range S_R of the sensor. The dimensions C_D of the coverage area are defined as a function of S_R such that a rectangular area with (height, width) $\leq C_D$ centered on a sensor covers all hotspots within that area with acceptable accuracy. Thus, coverage area dimension $C_D = \sqrt{2S_R}$. Although this type of grid-based sensor placement works reasonably well, the number of sensors required increases proportional to the whole CLB array size. In order to bound the sensor array size, optimized placement algorithms have been proposed [100–103].

One method utilized the recursive bisectioning concept borrowed from physical design. This idea is illustrated in Figure 2.28(b) and compared against the simpler grid-based approach. The goal is again to minimize the number of sensors and determine their placement to cover a set of hotspots. Each sensor has a coverage area dimension C_D as defined earlier. Sensor s_i can monitor hotspots that lie within the coverage area of height and width at most equal to C_D. This algorithm tries to create a set of rectangles each having their height and width at most C_D containing a set of hotspots. The sensor is placed at the center of such a rectangle. It is shown that the actual region formed can be smaller than this dimension depending on the location of the hotspots that the sensor covers. That is why the shape of the region is referred to as a rectangle as opposed to a square of maximum dimensions of $C_D \times C_D$. Let us assume that a bounding rectangle R is created around five hotspots $h_1, h_2, \ldots, h_5$. The sensor S is at the center of R. Let the distance of the hotspot h_1 from the sensor be r_1. Similarly the distances of other hotspots are denoted by $r_2, \ldots, r_5$. Since the sides of the bounding rectangle are at most C_D, the maximum distance of any hotspot within R can be S_R from S, if $C_D = 2.S_R$. This is shown Figure 2.29.

The hotspots h_1 and h_5 are at the two corners and farthest from sensor S. $r_1 = r_5 = S_R$ in this case. The other hotspots are closer to S, that is, $r_{2,3,4} < S_R$. Thus placing the sensor at the center of the bounding rectangle of height and width of at

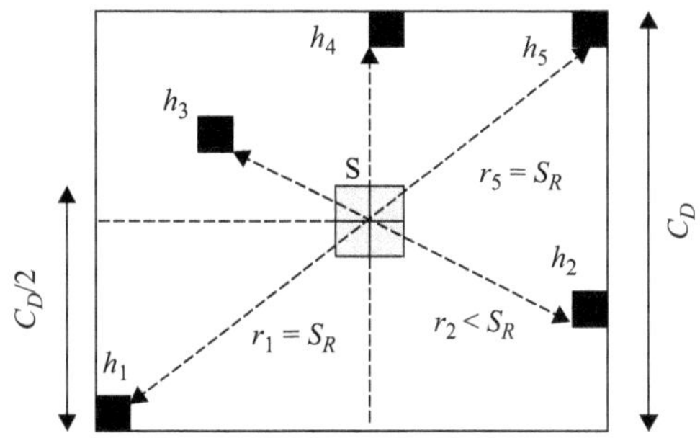

Figure 2.29 Sensor at the center of coverage area of dimension $C_D \times C_D$

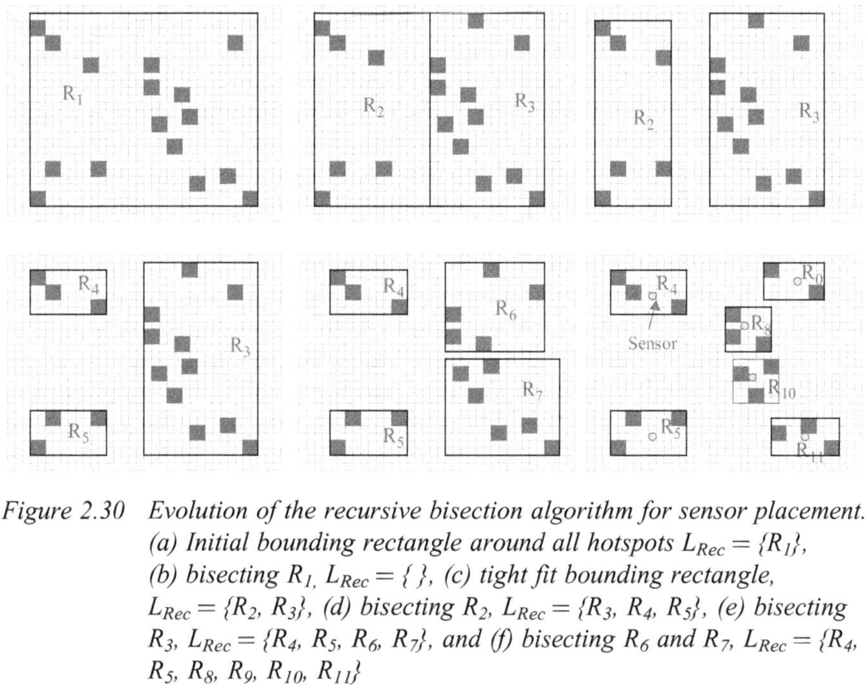

Figure 2.30 Evolution of the recursive bisection algorithm for sensor placement. (a) Initial bounding rectangle around all hotspots $L_{Rec} = \{R_1\}$, (b) bisecting R_1, $L_{Rec} = \{\}$, (c) tight fit bounding rectangle, $L_{Rec} = \{R_2, R_3\}$, (d) bisecting R_2, $L_{Rec} = \{R_3, R_4, R_5\}$, (e) bisecting R_3, $L_{Rec} = \{R_4, R_5, R_6, R_7\}$, and (f) bisecting R_6 and R_7, $L_{Rec} = \{R_4, R_5, R_8, R_9, R_{10}, R_{11}\}$

most C_D guarantees that all hotspots within this rectangle are within the range of the sensor.

The algorithm works by recursively creating a bounding rectangle R_i around hotspots and bisecting the rectangle along an edge until both height and width of R_i is less than or equal to C_D. The number of sensors n corresponds to the number of rectangles $R_p, \ldots, R_q$. Each R_i has a list of hotspots L_H^i that its perimeter covers. Two pairs of (x, y) coordinates can be used to represent R_i, which we denote by $[(x_{left}^i, y_{bottom}^i), (x_{right}^i, y_{top}^i)]$. The steps of the algorithm are illustrated in Figure 2.30(a–f).

2.6.4 Recent developments in sensor allocation and placement

The early works on temperature sensor allocation and placement for processors and reconfigurable devices were followed by approaches to obtain refined accuracies in monitoring and/or to minimize the hardware cost [104–109].

One approach formulates the problem in a form of matching between profiled hotspot locations in a grid and a set of candidate locations, where the layout and other design constraints allow placement of temperature sensors. The set of hotspot locations are given with the set $Q = [q_1, \ldots, q_n]$ with associated accuracy expectations for each location given as $E = [e_1, \ldots, e_n]$. Although it is unlikely that a separate accuracy for monitoring each such individual point would be required (or even could be quantified), the general formulation of the problem might still be insightful and it can always be reduced. Similarly, the set of candidate sensor locations is $L = [l_1, \ldots, l_k]$. The objective of this method is defined as finding the minimum subset S of L, such that, for each hotspot q_i, there is a matched sensor s_j, such that at all times the difference between the temperature reported by the sensor s_j and the temperature of the point q_i is less than the accuracy bound e_i [104]. This problem is solved by defining an observable area (according to a distance-thermal monitoring accuracy model) around each sensor for a given accuracy. This model follows a similar rationale to the circular regions defined around hotspots discussed in the earlier section [100]. Any hotspot location on those grid points that fall within the observable area of a sensor is a match. An integer linear programming (ILP)-based formulation is used to solve this optimization problem, since the problem has been converted to a discrete selection problem via the use of grid locations.

Another sensor allocation technique argues that more emphasis should be placed on minimizing the worst-case temperature tracking error rather than the average monitoring error, since the highest peak points will dictate the frequency of DTM interventions and affect reliability more directly [105]. To attack this problem, a two-phase heuristic approach has been proposed. The first phase produces an initial set of locations, and the second phase improves upon the results of the first phase in an iterative fashion. The first phase has been implemented with a variant of matrix volume sampling technique [110]. The main idea is to start with p potential sensor locations and create an initial solution with k sensors representing locations with maximally diverse (independent) thermal behavior. Each sensor is represented with a vector of n temperature values derived from a trace of thermal behavior observed at that location through simulations. The entire sensor set is then represented with a $p \times n$ matrix. This phase starts out by first selecting a sensor location where the L2 norm of temperature is highest. In subsequent $(k - 1)$ iterations, the algorithm computes the orthogonal components of the column vectors at the available sensors locations and picks the location with highest orthogonal L2 norm. Starting with this initial set of k sensors, the second phase of the algorithm iterates once over each sensor in this set. At each iteration, the locations of the $k - 1$ sensors are fixed and the best possible location for the one sensor under consideration is found to achieve the minimum worst-case monitoring error.

A recent study on sensor placement first aims to solve the problem of reconstructing a detailed thermal map from a limited set of sensors. Observations and computational techniques developed for this task are then re-applied to sensor placement problem, such that a good sensor placement is defined to be one that minimizes the thermal map reconstruction error [107].

2.7 Indirect temperature sensing

While the fields of on-chip temperature sensor design and optimizations for sensor allocation and placement have been thriving, there were also attempts at inferring the thermal state of an IC via alternative indicators. These approaches are mainly developed for microprocessor chips, where the software layer has access to dynamically collected statistics of system events and performance indicators. Light-weight online temperature predictors that purely rely on these indicators have been investigated. Performing a detailed thermal simulation of the underlying system using a thermal network model similar to the one presented in Figure 1.9 in Chapter 1 is infeasible at run-time in most practical scenarios. However, various proposals considered predicting temperature through simplified relationships, which capture the thermal behavior in some analytical form as a function of activity or estimated power consumption per system event.

It is obvious from the trends of including increasing numbers on on-chip sensors in several generations of commercial microprocessor products that indirect methods are unlikely to replace on-chip sensors. Quite to the contrary, embedded sensors are proliferating in many other system components (e.g., DRAMs, batteries in mobile devices). Nevertheless, indirect means can provide useful insights and cooperate with hardware-based thermal monitoring units.

Software-based thermal monitors usually are based on the concept of devising a very fast method for solving equivalent thermal network equations using estimated power from performance counters as input. They mainly differ in the specific computational tools they use to perform this ultra-fast thermal simulation at run-time. In one method, software-level implementation of light-weight real-time indicators of temperature have been called virtual sensors [83], where a prior dynamic online power model [111] has been coupled with a fast solver of a thermal RC-network model. Another method proposes the implementation of so-called software sensors, where performance counter statistics are fed into a numerical method to solve for temperature [112]. Power consumption of a processor core (Pentium 4 in this case) is estimated online by associating per access power of a functional unit with the total activity obtained from performance counters. The thermal equivalent second-order differential equations describing the thermal response of the core is then solved with the input power through a simplification. Specifically, moment matching in the frequency domain is adopted to compute the transient thermal response of the system. The main assumption is that the equivalent thermal model is dominated by a few system states and their corresponding poles in the transfer function can be found via moment matching. A specialized solver has also been proposed based on the linear system theory to model a chip as a time-invariant linear system with thermal response linearly dependent on power [113].

Indirect temperature tracking should be viewed as a distinct sub-domain from predictive methods that aim to guess the thermal state of a system at a distant future. In this specific context, we refer only to those schemes that estimate the current thermal state based on the current activity indicators of a system. These indirect tracking methods generally do not involve any feedback from on-chip and on-board hardware monitors of temperature and current (used to estimate power). They are truly software-based in the sense that they rely on performance counters accessible to the operating system (OS) and workload and utilization indicators. On the other hand, prediction schemes usually involve a larger window of history on activity indicators as well as feedback from hardware monitors including temperature sensors.

References

[1] Souri, K., Y. Chae, and K.A.A. Makinwa, A CMOS temperature sensor with a voltage-calibrated inaccuracy of 0.15 C (3) from 55 C to 125 C. IEEE Journal of Solid-State Circuits, 2013. **48**(1): pp. 292–301.

[2] McGowen, R., et al., Power and temperature control on a 90-nm Itanium family processor. IEEE Journal of Solid-State Circuits, 2006. **41**(1): pp. 229–237.

[3] Shor, J. and K. Luria, Evolution of thermal sensors in Intel processors from 90nm to 22nm. in Electrical & Electronics Engineers in Israel (IEEEI), 2012 IEEE 27th Convention of. 2012. IEEE.

[4] Mesa-Martinez, F.J., E.K. Ardestani, and J. Renau, *Characterizing processor thermal behavior*. in *ACM SIGARCH Computer Architecture News*. 2010. ACM.

[5] Zhao, J., R. Tessier, and W. Burleson, *Distributed sensor data processing for many-cores*. in *Proceedings of the Great Lakes Symposium on VLSI*. 2012. ACM.

[6] Shor, J., Compact Thermal Sensors in Intel Processors from 90nm to 22nm February 17, 2014: Columbia University.

[7] Bakker, A., *CMOS smart temperature sensors-an overview*. in *Sensors, 2002. Proceedings of IEEE*. 2002. IEEE.

[8] Banba, H., et al. A CMOS bandgap reference circuit with sub-1-V operation. IEEE Journal of Solid-State Circuits, 1999. **34**(5): pp. 670–674.

[9] Verster, T.C., P–N junction as an ultralinear calculable thermometer. Electronics Letters, 1968. **4**(9): pp. 175–176.

[10] Pertijs, M.A.P., K.A.A. Makinwa, and J.H. Huijsing, A CMOS smart temperature sensor with a 3σ inaccuracy of ± 0.1 C from –55 C to 125 C. IEEE Journal of Solid-State Circuits, 2005. **40**(12): pp. 2805–2815.

[11] Duarte, D.E., et al., *Temperature sensor design in a high volume manufacturing 65nm CMOS digital process*. in *Custom Integrated Circuits Conference, 2007. CICC'07. IEEE*. 2007. IEEE.

[12] Ma, M., et al., *Enhanced thermal management for future processors.* in *VLSI Circuits, 2003. Digest of Technical Papers. 2003 Symposium on.* 2003. IEEE.

[13] Sebastiano, F., et al., A 1.2-V 10-W NPN-based temperature sensor in 65-nm CMOS with an inaccuracy of 0.2 C (3) from 70 C to 125 C. IEEE Journal of Solid-State Circuits, 2010. **45**(12): pp. 2591–2601.

[14] Li, Y.W., et al., *A 1.05 V 1.6 mW 0.45 C 3σ-resolution ΔΣ-based temperature sensor with parasitic-resistance compensation in 32nm CMOS.* in *Solid-State Circuits Conference-Digest of Technical Papers, 2009. ISSCC 2009. IEEE International.* 2009. IEEE.

[15] Sanchez, H., et al., A CMOS temperature sensor for PowerPC(TM)RISC microprocessors. 1997 Symposium on VLSI Circuits. 1997. pp. 13–14.

[16] Yoshida, M. and D.W. Boerstler, *Thermal sensing circuit using bandgap voltage reference generators without trimming circuitry.* 2010. Google Patents, US Patent US 7789558 B2.

[17] Floyd, M., et al., Introducing the adaptive energy management features of the Power7 chip. IEEE Micro, 2011. **31**(2): pp. 60–74.

[18] Lee, T.H., M.G. Johnson, and M.P. Crowley, *Temperature sensor integral with microprocessor and methods of using same.* 1999. Google Patents, US Patent US 5961215 A.

[19] Chowdhury, G. and A. Hassibi, An on-chip temperature sensor with a self-discharging diode in 32-nm SOI CMOS. Circuits and Systems II: Express Briefs, IEEE Transactions on, 2012. **59**(9): pp. 568–572.

[20] Cahoon, C. and R.J. Baker, *Low-voltage CMOS temperature sensor design using schottky diode-based references.* in *Microelectronics and Electron Devices, 2008. WMED 2008. IEEE Workshop on.* 2008. IEEE.

[21] Floyd, M.S., et al., System power management support in the IBM POWER6 microprocessor. IBM Journal of Research and Development, 2007. **51**(6): pp. 733–746.

[22] Datta, B. and W.P. Burleson, *Low power on-chip thermal sensors based on wires.* in *Very Large Scale Integration, 2007. VLSI-SoC 2007. IFIP International Conference on.* 2007. IEEE.

[23] Kasap, S., Thermoelectric Effects in Metals: Thermocouples. Canada: Department of Electrical Engineering. 2001, University of Saskatchewan. Saskatchewan, Canada.

[24] Awtrey, D., *The 1-Wire Thermocouple.* http://www.1wire.org/Files/Awtrey/Articles/1-Wire thermocouple.doc, 2002.

[25] Herwaarden, A.W.V. and P.M. Sarro, Thermal sensors based on the Seebeck effect. Sensors and Actuators, 1986. **10**(1986): pp. 321–346.

[26] Khadouri, S., G.C.M. Meijer, and F.M.L. Van der Goes, A CMOS Interface for Thermocouples with Reference-Junction Compensation. *Analog Integrated Circuits and Signal Processing.* 1997. **14**: pp. 235–248, Kluwer Academic Publishers, Boston, MA, USA.

[27] Mitcheli, M., *Implementing a Single-Chip Thermocouple Interface with the MSP430x42x.* focus.tij.co.jp/jp/lit/an/slaa216/slaa216.pdf, 2005.

[28] Baglio, S., S. Castorina, and N. Savalli, *On-chip temperature monitoring via CMOS thermocouples*. in *THERMINIC*. 2003.
[29] Liu, H., et al., Thin-film thermocouple array for time-resolved local temperature mapping. Electron Device Letters, IEEE, 2011. **32**(11): pp. 1606–1608.
[30] Park, J.-J. and M. Taya, Design of micro-temperature sensor array with thin film thermocouples. Journal of Electronic Packaging, 2005. **127**(3): pp. 286–289.
[31] Li, D., J.-H. Kim, and S.O. Memik, *Integrating thermocouple sensors into 3D ICs*. in *Computer Design (ICCD), 2013 IEEE 31st International Conference on*. 2013. IEEE.
[32] Grayson, M., et al., *Thin film thermocouples for thermoelectric characterization of nanostructured materials*. in *APS Meeting Abstracts*. 2011, Dallas, TX, USA.
[33] Liu, H., et al., Towards on-chip time-resolved thermal mapping with micro-/nanosensor arrays. Nanoscale Research Letters, 2012. **7**(1): pp. 1–6.
[34] Varrenti, A.R., et al., Thermal sensing with lithographically patterned bimetallic thin-film thermocouples. Electron Device Letters, IEEE, 2011. **32** (6): pp. 818–820.
[35] Long, J., et al., *IOTA: Towards an integrated on-chip thermocouple array*. in *Thermal Investigations of ICs and Systems (THERMINIC), 2010 16th International Workshop on*. 2010. IEEE.
[36] Grayson, M.A., et al., *Bimetallic Integrated On-Chip Thermocouple array*. 2013. Google Patents.
[37] van Vroonhoven, C.P.L. and K.A.A. Makinwa, An SOI thermal-diffusivity-based temperature sensor with $\pm 0.6^\circ$ C (3σ) untrimmed inaccuracy from$-$70° C to 225° C. Sensors and Actuators A: Physical, 2012. **188**: pp. 66–74.
[38] van Vroonhoven, C.P.L. and K.A.A. Makinwa, *Thermal diffusivity sensing: A new temperature sensing paradigm*. in *Custom Integrated Circuits Conference (CICC), 2011 IEEE*. 2011. IEEE.
[39] Vittoz, E. and J. Fellrath, *CMOS analog integrated circuits based on weak inversion operations*. IEEE Journal of Solid-State Circuits, 1977. **12**(3): pp. 224–231.
[40] Schneider, M.C. and C. Galup-Montoro, *CMOS Analog Design Using All-Region MOSFET Modeling*. 2010. Cambridge University Press, Cambridge, UK.
[41] Datta, B., *On-Chip Thermal Sensing in Deep Sub-micron CMOS*. 2007. University of Massachusetts, Amherst, MA, USA.
[42] Meijer, G.C.M., G. Wang, and F. Fruett, Temperature sensors and voltage references implemented in CMOS technology. IEEE Sensors Journal, 2001. **1**(3): pp. 225–234.
[43] Dantas, J.M.C., et al., *Low-power high-rensponsivity CMOS temperature sensor*. in *Instrumentation and Measurement Technology Conference Proceedings, 2008. IMTC 2008. IEEE*. 2008. IEEE.
[44] Sasaki, M., M. Ikeda, and K. Asada, A temperature sensor with an inaccuracy of C using 90-nm 1-V CMOS for online thermal monitoring of VLSI

circuits. Semiconductor Manufacturing, IEEE Transactions on, 2008. **21**(2): pp. 201–208.
[45] Law, M.K. and A. Bermak, A 405-nW CMOS temperature sensor based on linear MOS operation. Circuits and Systems II: Express Briefs, IEEE Transactions on, 2009. **56**(12): pp. 891–895.
[46] Székely, V., et al., CMOS sensors for on-line thermal monitoring of VLSI circuits. Very Large Scale Integration (VLSI) Systems, IEEE Transactions on, 1997. **5**(3): pp. 270–276.
[47] Miribel-Català, P., et al., MOSFET-based temperature sensor for standard BCD smart power technology. Microelectronics Journal, 2001. **32**(10): pp. 869–873.
[48] Jeong, Y. and F. Ayazi, *Process compensated CMOS temperature sensor for microprocessor application.* in *Circuits and Systems (ISCAS), 2012 IEEE International Symposium on*. 2012. IEEE.
[49] Berber, F., *CMOS Temperature Sensor Utilizing Interface-Trap Charge Pumping*. 2006. Texas A&M University, College Station, TX, USA.
[50] Chen, P., et al., A time-to-digital-converter-based CMOS smart temperature sensor. IEEE Journal of Solid-State Circuits, 2005. **40**(8): pp. 1642–1648.
[51] Chen, P., et al., *A time-domain SAR smart temperature sensor with* $-0.25 \sim +0.35°C$ *inaccuracy for on-chip monitoring.* in *Solid-State Circuits Conference, 2008. ESSCIRC 2008. 34th European*. 2008. IEEE.
[52] Chen, P., et al., A time-domain SAR smart temperature sensor with curvature compensation and a 3σ inaccuracy of -0.4 C $\sim +0.6$ C over a 0 C to 90 C range. IEEE Journal of Solid-State Circuits, 2010. **45**(3): pp. 600–609.
[53] Ha, D., et al., Time-domain CMOS temperature sensors with dual delay-locked loops for microprocessor thermal monitoring.Very Large Scale Integration (VLSI) Systems, IEEE Transactions on, 2012. **20**(9): pp. 1590–1601.
[54] Kim, K., et al., *A 366kS/s 400uW 0.0013 mm 2 frequency-to-digital converter based CMOS temperature sensor utilizing multiphase clock.* in *Custom Integrated Circuits Conference, 2009. CICC'09. IEEE.* 2009. IEEE.
[55] Chen, S.-W., et al., *Fully on-chip temperature, process, and voltage sensors.* in *Circuits and Systems (ISCAS), Proceedings of 2010 IEEE International Symposium on*. 2010. IEEE.
[56] Xie, S. and W.T. Ng., *Delay-line based temperature sensors for on-chip thermal management.* in *Solid-State and Integrated Circuit Technology (ICSICT), 2012 IEEE 11th International Conference on*. 2012. IEEE.
[57] Xie, S. and W.T. Ng., *A 0.02 nJ self-calibrated 65nm CMOS delay line temperature sensor.* in *Circuits and Systems (ISCAS)*, 2012 *IEEE International Symposium on.* 2012. IEEE.
[58] Chiang, T.-T., et al., *On-chip self-calibrated process-temperature sensor for TSV 3D integration*. in *SOC Conference (SOCC), 2012 IEEE International.* 2012. IEEE.
[59] Kashfi, F. and J. Draper., *Thermal sensor design for 3D ICs.* in *Circuits and Systems (MWSCAS), 2012 IEEE 55th International Midwest Symposium on.* 2012. IEEE.

[60] Boemo, E. and S. López-Buedo, *Thermal monitoring on FPGAs using ring-oscillators*. in *Field-Programmable Logic and Applications*. 1997. Springer.
[61] Kim, C.-K., et al., CMOS temperature sensor with ring oscillator for mobile DRAM self-refresh control. Microelectronics Journal, 2007. **38**(10): pp. 1042–1049.
[62] Franco, J.J.L., et al., *Ring oscillators as thermal sensors in FPGAs: Experiments in low voltage*. in *Programmable Logic Conference (SPL), 2010 VI Southern*. 2010. IEEE.
[63] Chen, P., et al., All-digital time-domain smart temperature sensor with an inter-batch inaccuracy of after one-point calibration. Circuits and Systems I: Regular Papers, IEEE Transactions on, 2011. **58**(5): pp. 913–920.
[64] Wolpert, D. and P. Ampadu, *Managing Temperature Effects in Nanoscale Adaptive Systems*. 2011. Springer Science & Business Media, New York, USA.
[65] Kaxiras, S. and P. Xekalakis, *4T-decay sensors: A new class of small, fast, robust, and low-power, temperature/leakage sensors*. in *Low Power Electronics and Design*, 2004. *ISLPED'04. Proceedings of the 2004 International Symposium on*. 2004. IEEE.
[66] Ituero, P., J.L. Ayala, and M. Lopez-Vallejo, *Leakage-based on-chip thermal sensor for CMOS technology*. in *Circuits and Systems, 2007. ISCAS 2007. IEEE International Symposium on*. 2007. IEEE.
[67] Ituero, P., J.L. Ayala, and M. Lopez-Vallejo, A nanowatt smart temperature sensor for dynamic thermal management. Sensors Journal, IEEE, 2008. **8**(12): pp. 2036–2043.
[68] Ituero, P., M. López-Vallejo, and C. López-Barrio, A 0.0016 mm2 0.64 nJ leakage-based CMOS temperature sensor. Sensors, 2013. **13**(9): pp. 12648–12662.
[69] Van de Plassche, R.J., *CMOS integrated analog-to-digital and digital-to-analog converters*. **Vol. 2**. 2003. Kluwer Academic Publishers Dordrecht, Boston, MA, USA.
[70] Bakker, A. and J.H. Huijsing, Micropower CMOS temperature sensor with digital output. IEEE Journal of Solid-State Circuits, 1996. **31**(7): pp. 933–937.
[71] Riedijk, F.R. and J.H. Huijsing, An integrated absolute temperature sensor with sigma-delta A–D conversion. Sensors and Actuators A: Physical, 1992. **34**(3): pp. 249–256.
[72] Wu, C.-K., W.-S. Chan, and T.-H. Lin, *A 80kS/s 36μW resistor-based temperature sensor using BGR-free SAR ADC with a unevenly-weighted resistor string* in *0.18μm CMOS*. in *VLSI Circuits (VLSIC), 2011 Symposium on*. 2011. IEEE.
[73] Chen, P., S.-I. Liu, and J. Wu, A CMOS pulse-shrinking delay element for time interval measurement. Circuits and Systems II: Analog and Digital Signal Processing, IEEE Transactions on, 2000. **47**(9): pp. 954–958.

[74] Chen, C.-C. and H.-W. Chen, A linearization time-domain CMOS smart temperature sensor using a curvature compensation oscillator. Sensors, 2013. **13**(9): pp. 11439–11452.

[75] Pertijs, M.A.P. and J.H. Huijsing, *Precision temperature sensors in CMOS technology*. 2006. Springer Science & Business Media, The Netherlands.

[76] Pertijs, M.A.P., A. Bakker, and J.H. Huijsing, *A high-accuracy temperature sensor with second-order curvature correction and digital bus interface*. in *Circuits and Systems, 2001. ISCAS 2001. The 2001 IEEE International Symposium on*. 2001. IEEE.

[77] Gunawan, M., et al., A curvature-corrected low-voltage bandgap reference. IEEE Journal of Solid-State Circuits, 1993. **28**(6): pp. 667–670.

[78] Long, J. and S.O. Memik, *Optimization of the bias current network for accurate on-chip thermal monitoring*. in *Design, Automation & Test in Europe Conference & Exhibition (DATE), 2010*. 2010. IEEE.

[79] Solbrekken, G.L. and C.-P. Chiu, Calibration of resistance type die level temperature sensors using a single temperature technique. Components and Packaging Technologies, IEEE Transactions on, 2000. **23**(1): pp. 40–46.

[80] Woo, K., et al., *Dual-DLL-based CMOS all-digital temperature sensor for microprocessor thermal monitoring*. in *Solid-State Circuits Conference-Digest of Technical Papers, 2009. ISSCC 2009. IEEE International*. 2009. IEEE.

[81] Ueno, K., T. Asai, and Y. Amemiya, Low-power temperature-to-frequency converter consisting of subthreshold CMOS circuits for integrated smart temperature sensors. Sensors and Actuators A: Physical, 2011. **165**(1): pp. 132–137.

[82] Aita, A.L., et al., *A CMOS smart temperature sensor with a batch-calibrated inaccuracy of* ± 0.25 *C* (3σ) *from* -70 *C to 130 C*. in *Solid-State Circuits Conference-Digest of Technical Papers, 2009. ISSCC 2009. IEEE International*. 2009. IEEE.

[83] Lee, K.-J., K. Skadron, and W. Huang, *Analytical model for sensor placement on microprocessors*. in *Computer Design: VLSI in Computers and Processors, 2005. ICCD 2005. Proceedings. 2005 IEEE International Conference on*. 2005. IEEE.

[84] Henning, J.L., SPEC CPU2000: Measuring CPU performance in the new millennium. Computer, 2000. **33**(7): pp. 28–35.

[85] Skadron, K., et al., *Temperature-aware microarchitecture*. in *ACM SIGARCH Computer Architecture News*. 2003. ACM.

[86] Mukherjee, R. and S.O. Memik, Systematic temperature sensor allocation and placement for microprocessors. 43rd Design Automation Conference, Proceedings 2006, **2006**: pp. 542–547.

[87] Long, J., et al., Thermal monitoring mechanisms for chip multiprocessors. ACM Transactions on Architecture and Code Optimization, 2008. **5**(2): pp. 9:1–9:33.

[88] Jain, A.K., Data clustering: 50 years beyond K-means. *Pattern Recognition Letters*, 2010. **31**(8): pp. 651–666.

[89] Hartigan, J.A. and M.A. Wong, Algorithm AS 136: A k-means clustering algorithm. Applied Statistics, 1979. **28**(1): pp. 100–108.

[90] Moore, A., K-means and Hierarchical Clustering. 2001. Carnegie Mellon University, Pittsburgh, PA, USA, http://www.autonlab.org/tutorials/kmeans.html

[91] Memik, S.O., et al., Optimizing thermal sensor allocation for microprocessors. IEEE Transactions on Computer-Aided Design of Integrated Circuits and Systems, 2008. **27**(3): pp. 516–527.

[92] John, J.K., J.S. Hu, and S.G. Ziavras, *Optimizing the thermal behavior of subarrayed data caches.* in *Computer Design: VLSI in Computers and Processors, 2005. ICCD 2005. Proceedings. 2005 IEEE International Conference on*. 2005. IEEE.

[93] Kaxiras, S., Z. Hu, and M. Martonosi, Cache decay: Exploiting generational behavior to reduce cache leakage power. ACM SIGARCH Computer Architecture News, 2001. **29**(2): pp. 240–251.

[94] Ku, J.C., et al., *Thermal management of on-chip caches through power density minimization.* in *Proceedings of the 38th annual IEEE/ACM International Symposium on Microarchitecture*. 2005. IEEE Computer Society.

[95] Binkert, N., et al., The gem5 simulator. ACM SIGARCH Computer Architecture News, 2011. **39**(2): pp. 1–7.

[96] Brooks, D., V. Tiwari, and M. Martonosi, Wattch: A Framework for Architectural-Level Power Analysis and Optimizations. *ACM SIGARCH Computer Architecture News – Special Issue: Proceedings of the 27th Annual International Symposium on Computer Architecture*, 2000. **28**(2): pp. 83–94.

[97] Huang, W., et al., HotSpot: A compact thermal modeling methodology for early-stage VLSI design. Very Large Scale Integration (VLSI) Systems, IEEE Transactions on, 2006. **14**(5): pp. 501–513.

[98] Lesea, A. and M. Alexander, *Powering xilinx fpgas. Application Notes, 2002.*

[99] Lopez-Buedo, S., J. Garrido, and E. Boemo, Dynamically inserting, operating, and eliminating thermal sensors of FPGA-based systems. Components and Packaging Technologies, IEEE Transactions on, 2002. **25**(4): pp. 561–566.

[100] Mondal, S., R. Mukherjee, and S.O. Memik, Fine-grain thermal profiling and sensor insertion for FPGAs. 2006 IEEE International Symposium on Circuits and Systems, vols 1–11, Proceedings, 2006: pp. 4387–4390.

[101] Mukherjee, R., S. Mondal, and S.O. Memik, Thermal sensor allocation and placement for reconfigurable systems. IEEE/Acm International Conference on Computer-Aided Design, Digest of Technical Papers, Iccad, 2006: pp. 605–610.

[102] Mukherjee, R., S. Mondal, and S.O. Memik, *A sensor distribution algorithm for FPGAs with minimal dynamic reconfiguration overhead*. in *Ersa*. 2006.

[103] Byunghyun, L., et al. Thermal sensor allocation and placement for reconfigurable systems. ACM Transactions on Design Automation of Electronic Systems, 2009. **14**(4): pp. 1–23.
[104] Sharifi, S. and T.S. Rosing, Accurate direct and indirect on-chip temperature sensing for efficient dynamic thermal management. Computer-Aided Design of Integrated Circuits and Systems, IEEE Transactions on, 2010. **29** (10): pp. 1586–1599.
[105] Reda, S., R.J. Cochran, and A.N. Nowroz, Improved thermal tracking for processors using hard and soft sensor allocation techniques. Computers, IEEE Transactions on, 2011. **60**(6): pp. 841–851.
[106] Zanini, F., et al., *Temperature sensor placement in thermal management systems for MPSoCs*. in *Circuits and Systems (ISCAS), Proceedings of 2010 IEEE International Symposium on*. 2010. IEEE.
[107] Ranieri, J., et al., *EigenMaps: Algorithms for optimal thermal maps extraction and sensor placement on multicore processors*. in *Design Automation Conference (DAC), 2012 49th ACM/EDAC/IEEE*. 2012. IEEE.
[108] Nowroz, A.N., R. Cochran, and S. Reda. *Thermal monitoring of real processors: Techniques for sensor allocation and full characterization*. in *Proceedings of the 47th Design Automation Conference*. 2010. ACM.
[109] Ituero, P., F. Garcia-Redondo, and M. Lopez-Vallejo, *Temperature sensor placement including routing overhead and sampling inaccuracies*. in *Synthesis, Modeling, Analysis and Simulation Methods and Applications to Circuit Design (SMACD), 2012 International Conference on*. 2012. IEEE.
[110] Çivril, A. and M. Magdon-Ismail, On selecting a maximum volume submatrix of a matrix and related problems. Theoretical Computer Science, 2009. **410**(47): pp. 4801–4811.
[111] Isci, C. and M. Martonosi, *Runtime power monitoring in high-end processors: Methodology and empirical data*. in *Proceedings of the 36th Annual IEEE/ACM International Symposium on Microarchitecture*. 2003. IEEE Computer Society.
[112] Wei, W., et al., Efficient power modeling and software thermal sensing for runtime temperature monitoring. ACM Transactions on Design Automation of Electronic Systems, 2008. **12**(3): pp. 1–29.
[113] Han, Y., I. Koren, and C.M. Krishna. *Temptor: A lightweight runtime temperature monitoring tool using performance counters*. in *Workshop on TACS*. 2006.

Chapter 3
Dynamic thermal management

A wide variety of processor and DRAM chips and system-on-chip (SoC) modules in the market today employ mechanisms to manage supply voltage, clock frequency, and cooling based on the feedback received from on-chip temperature sensors. Most commercial realizations involve an additional on-chip thermal management unit: for example, the on-chip thermal assist unit (TAU) in early PowerPC processors [1], EnergyScale microcontroller in IBM Power6 and Power7 [2–4], Intel's Thermal Monitor TM1 and TM2 technologies [5], and AMD's on-chip thermal evaluation centers [6]. In addition, there is a large body of academic work on enabling proactive and reactive thermal management at run-time. These schemes are broadly called DTM. In most cases, the DTM schemes coordinate closely with the dynamic power management mechanisms as well. Hence, in this chapter the two mechanisms will be described together in some cases.

On-chip temperature sensors are essential components of DTM units deployed in high-performance chips and high-end small form factor mobile devices. DTM has become a rich field of industrial innovation and academic research in the last decade. In order to maximize the coverage, an increasing number of on-chip thermal sensing devices are scattered across the entire chip. They are networked by an underlying infrastructure, which provides the bias currents/voltages to the sensing devices, collects measurements, performs A/D signal conversion, and delivers the temperature readings to an external thermal management unit. It needs to span a large distance covering the entire processor core, networking an increasing number of devices. Therefore, the supporting infrastructure is an on-chip element at a global scale, growing in complexity with each emerging processor generation and its design is now a non-trivial task. Furthermore, the specific ways in which the thermal management system reacts to sensor readings to prevent excessive thermal stress is an important problem. Different systems accomplish this in various ways and a large body of academic work has been devoted to alternative methods.

In this chapter, we will first describe prominent examples of the global thermal management infrastructures, including sensor interfaces and feedback control pathways. Next, methods for DTM involving the interaction between the sensing, control, and intervention will be discussed. To a first degree, clock frequency and supply voltage are managed to control the hardware's thermal response. This management occurs at two major levels of abstraction. A majority of first response mechanisms are carried out directly by the hardware. Temperature and power measurements are also reported to the system software through various interface technologies. Combining this information reported by the hardware with system-level knowledge of parameters such as processor utilization, the system software can also perform additional management, usually at a larger timescale.

We will discuss algorithms that these global DTM schemes execute to control the underlying system's thermal state by manipulating the major performance parameters (e.g., voltage, frequency) using temperature measurement feedback. We will also discuss academic studies that aim to take the system-level thermal management one step further through manipulation of additional decision mechanisms such task assignment, data layout (for memory chips), and task migration.

3.1 Interfacing temperature sensors and DTM systems

The hardware structure of the DTM systems can be divided into two main components: the sensor network including the interconnect network carrying temperature sensor outputs and the control component interfacing with the voltage and frequency management units. Arguably the design of the overall thermal monitoring and management infrastructure is as important for the overall accuracy and quality of the thermal monitoring as the individual sensing devices contained within this network. Temperature sensors on one hand receive biasing currents and/or voltages from a network of sources and mirroring circuits. On the other hand, their digital outputs are routed to one or few central locations for processing by the DTM hardware. These structures will be described in the following sections.

3.1.1 Biasing networks for temperature sensors

Current distribution structures are an integral part of analog temperature sensors networks on chips. These sensors need to be biased with a constant current source and they derive multiple reference current and voltages from them internally. Providing the biasing locally with dedicated analog circuits on the same location as the sensing point involves more complications. In older process nodes, it was possible to design these analog parts at a larger scale with better immunity to process and thermal variations. Various compensation techniques for temperature and process variation can also be applied to the current and voltage sources to improve their sensitivity to these sources of variation [7–10]. Furthermore, systematic variations can be accounted for to some extent during calibration. However, with the scaling of the process technology, it becomes increasingly challenging to design these circuits with the same level of immunity at the spot of sensing. Increasing die temperatures in modern chips also further exacerbates the thermal sensitivity of analog components.

In modern high-performance chips, biasing for the sensor back-ends is generated at a central location with access to better controlled and predictable analog source. For instance, in the case of the thermal monitoring system for the AMD quad-core Opteron processor, each core contains a number of remote temperature sensors scattered across the core and the sensor readings are routed to a central thermal evaluation unit [6]. Duarte et al. discussed advanced thermal sensing techniques used in an Intel Pentium 4 processor [11]. Both local and remote sensors are employed in the processor. In these systems, sensor back-end components that generate reference voltage and/or current for bandgap circuits need to be biased with a dedicated network of current sources. Different methodologies for constructing this infrastructure can lead to drastically different monitoring outcomes. We will illustrate the consequences of the network's structure by contrasting the choice between two main measurement techniques: two-wire and four-wire measurement. The choice of either technique to sample the sensor's voltage output will impact the current biasing network of the sensor's back-end.

A common method adopted by industrial designs [11, 12] is the two-wire measurement depicted in Figure 3.1. Notice that the temperature correlated sensor output voltage V_{BE} is measured and processed a certain distance away from the sensor. In this figure, the sensor is depicted symbolically as a diode, but it is essentially built with either a BJT or MOSFET in a variety of ways described in Chapter 2.

While the abovementioned challenges are inherent to the structure of the biasing circuitry, there is another source of inaccuracy stemming solely from the structure of the current distribution network. The wire connecting the sensor and the voltage measurement module is associated with series resistance R_s, which is a function of the temperature profiles along the length of the wire.

$$R_s = R_{s0} \cdot \left(1 + \beta \cdot \int_L (T(x,y) - T_0)dl \bigg/ \int_L dl \right) \tag{3.1}$$

R_{s0} is the resistivity of the interconnect at room temperature T_0 (assumed to be 25°C), β is the temperature coefficient of resistance and the integration is performed along the wire. In (3.1), the ratio of the two integration terms can be interpreted as the average variation of temperature with respect to the nominal temperature along the interconnect, denoted by ΔT_{avg}.

The junction temperature provided by the analog sensor is ultimately related to the measured voltage difference V_{meas} and R_s by

$$T_d = \alpha(V_{meas} - I_C R_s) + \tau \tag{3.2}$$

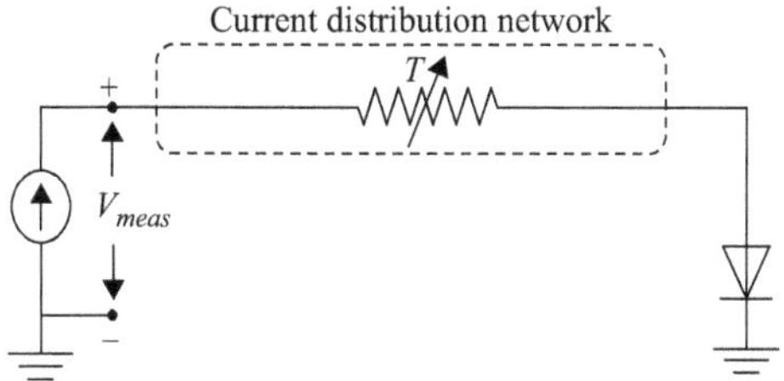

Figure 3.1 Two-wire voltage measurement from analog voltage output temperature sensors

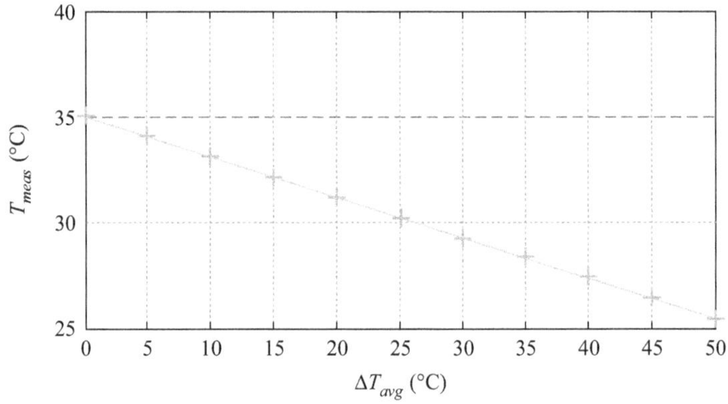

Figure 3.2 Interpretation of an analog sensor's output after transmission through a long interconnect as a function of ΔT_{avg}

Equations (3.1) and (3.2) indicate that without sufficient knowledge about the temperature distribution along the wire carrying the sensor back-end's output to the central sensor processing unit, the actual junction temperature cannot be precisely determined. The back-end and the measurement/digital-conversion circuitry can be placed far away from each other, such that the interconnect between them would have to cover significant distances. These lines may need to cross an entire processor core in some existing designs. It is certain that the thermal profile across such distances on a high-performance processor will vary significantly. Furthermore, a detailed map of this profile will not be available. Hence, the measurement error introduced by ΔT_{avg} cannot be compensated by any kind of calibration.

Figure 3.2 depicts an analysis on the impact of the interconnect's resistance on measurement's accuracy. This study was performed on 65 nm technology with copper interconnects with the width and length of the interconnect between the sensor back-end and voltage measurement point to be 180 nm and 3 mm, respectively. The bias current is 10 μA. The temperature at the sensor location is kept at 35°C and the temperature profile along the interconnect is varied. The figure plots the relationship between the measured temperature T_{meas} and ΔT_{avg} using a solid line. The measured temperature value is given by (3.2). The temperature at the sensor's location, which is kept constant at 35°C, is also depicted using a dashed horizontal line. As ΔT_{avg} increases, we observe that there exists a significant disagreement between the actual and measured temperature. For instance, when ΔT_{avg} is 25°C, the measurement error can reach 5°C. In modern microprocessors, when running different applications, the die temperature can be dramatically different at different locations. Depending on the thermal throttling threshold, critical functional units such as integer/floating ALU, integer/floating register file, and instruction queue can exceed 90°C. Temperatures of other functional units can be over 60°C. Hence, it is not unusual for ΔT_{avg} to be as large as 25°C.

In order to mitigate this source of inaccuracy, an alternative is to use the four-wire (Kelvin) measurement as depicted in Figure 3.3. In the four-wire

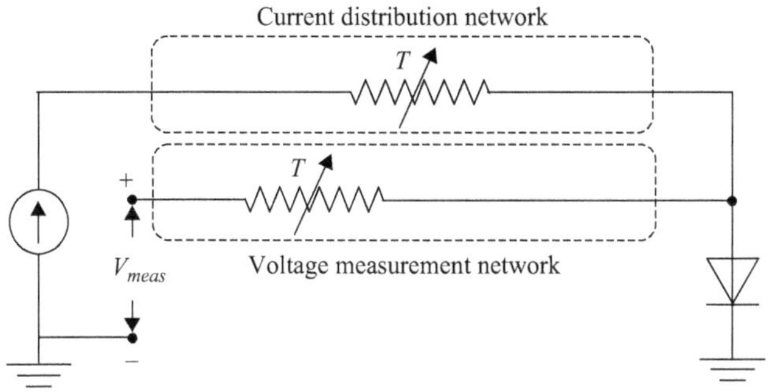

Figure 3.3 Four-wire (Kelvin) voltage measurement from the analog sensor back-end. The routing network contains a dedicated bias current distribution network and another dedicated voltage output collection network

measurement, since the electric current passing through the voltage measurement network is negligible, the measured voltage V_{meas} is equal to the sensor back-end's output voltage and hence, does not involve R_s. Therefore, the serial resistances of the wires and their thermal dependency do not impact the accuracy of the measurement. However, four-wire measurement is more resource-demanding, since the entire routing infrastructure contains the voltage measurement network in addition to the bias current distribution network. The cost of the current distribution network can be reduced by leveraging current mirrors.

Figure 3.4(a) depicts the structure of a cascode current mirror built using PMOS transistors. The cascode current mirror structure has high output resistance, which eliminates the coupling between the current mirrors and the thermal sensors. Figure 3.4(b) demonstrates the benefits of using a current mirror tree in comparison

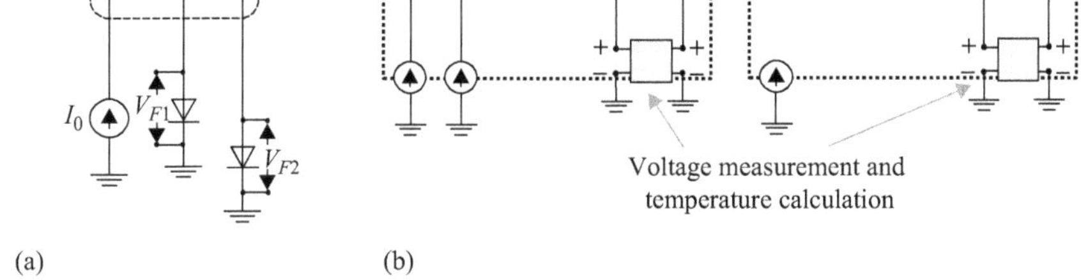

Figure 3.4 Leveraging the current mirror structure: (a) transistor-level schematic of the current distribution network and (b) the benefit obtained by introducing the current mirrors

to using a dedicated current source for each temperature sensor back-end on the chip [13]. One current source and part of the interconnect are eliminated after introducing a current mirror into the network. Notice that the resulting current distribution network has a tree topology.

The current mirrors could potentially be sources of inaccuracy to the monitoring system. First, due to process variation, the matching ratio of a current mirror (output current over input current) might not be exactly unity. As a result, the actual bias current of a thermal diode could deviate from the current generated by the current source. However, the impact of process variation on the system accuracy is static. In other words, the matching ratio of each individual current mirror is a fixed value after manufacturing. Therefore, the impact of process variation on sensor accuracy can be mitigated by sensor calibration. Second, spatial and temporal temperature variation affects the matching ratio of a current mirror as well. On the other hand, the interconnects for voltage measurement, would be directly routed from the thermal diode to the voltage measurement circuits as shown in Figure 3.5(b). Crosstalk-induced noise

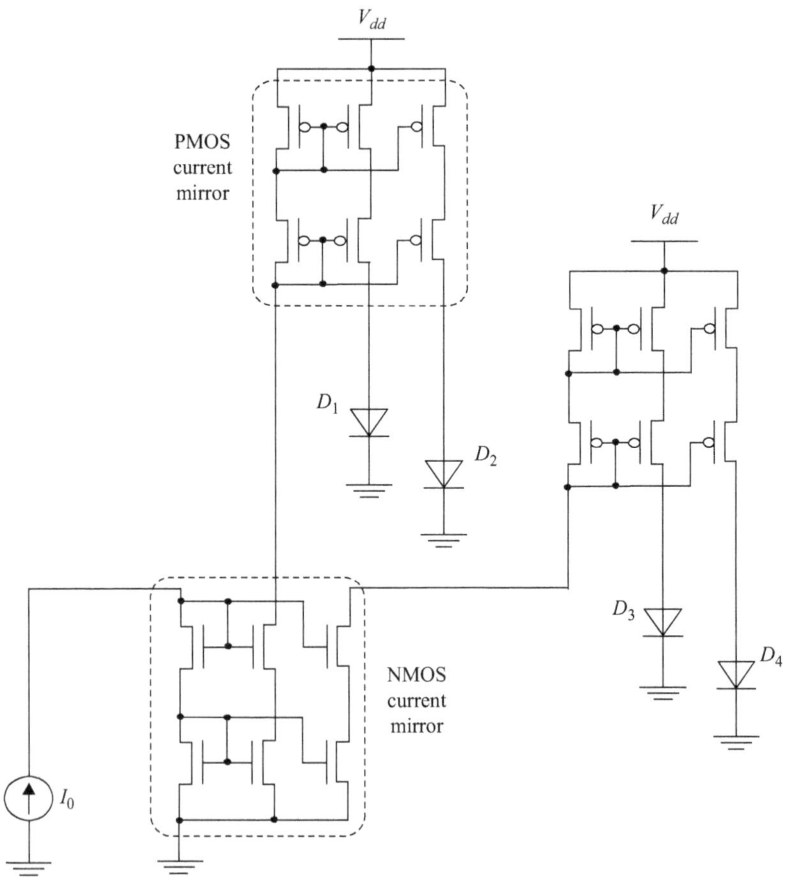

Figure 3.5 The bias current distribution network supporting four diodes

can be a source of inaccuracy for voltage measurement. However, this effect can be minimized by adding adequate shielding to the lines as it is currently practiced in industrial designs. Furthermore, the thermal time constant is in the order of milliseconds, which is several orders of magnitude larger than the clock cycle time (in the order of nanosecond). Thus, the voltage can be sampled in multiple clock cycles and the average voltage could also be used for temperature calculation. This would effectively filter out the inaccuracy imposed by the coupling between the interconnects.

This current network can also be extended to support a higher number of sensors. Figure 3.5 illustrates a network supporting four sensors. Once again, note that analog sensor back-ends have been depicted with a thermal diode for ease of illustration and they can be constructed with a variety of bandgap circuits. Notice that we use alternating NMOS and PMOS current mirrors in the tree. This is necessary in order to create current flows in the appropriate directions. The input/output currents of the NMOS current mirror flow inward, while the input/output currents of the PMOS current mirror flow outward.

The structure of the current mirror is quite simple. Available whitespace on the layout could be utilized to embed them into the chip. The area of a current mirror is comparable to that of a repeater. Therefore, like the repeaters, the current mirrors can be inserted into the whitespaces of the chips [14]. Hence, the insertion of current mirrors can be performed in post-layout stage. Therefore, the necessary modification made to the design flow is minimal. Furthermore, the current mirror can be extended to map one input current to more than two output currents. However, since the current mirrors are inserted into the whitespace whose capacity is limited, the number of output currents of a current mirror cannot be made arbitrarily large.

The routing network described above supports higher precision compared to the routing network used by the two-wire measurement. However, it has higher interconnect overhead. Current mirrors can help to reduce this overhead. However, the problem of allocation and placement of the current mirrors to obtain the minimum wire length remains to be addressed. A possible approach to tackle this challenge is to formulate the problem as a Steiner minimal tree problem on graphs. For a given chip, a weighted graph can be constructed, such that each analog sensor back-end/whitespace is represented by a vertex. The weight of an edge between two vertices equals to the Manhattan distance (assuming rectilinear routing) between the corresponding sensors/whitespaces. The minimal Steiner tree problem on graphs is a well-studied problem [15, 16]. However, this problem is distinct from other variants, because the Steiner points cannot have arbitrarily large degrees. As mentioned earlier, the capacity of each whitespace is limited. This means, the number of transistors, thereby the number of output currents of a current mirror occupying each whitespace cannot be made arbitrarily large. Therefore, each possible Steiner point is associated with an upper bound on the maximum degree on the Steiner tree. Also, the sensors constitute the leaf nodes of the tree. This constraint is equivalent to requiring the maximum degree of the sensor nodes to be one. Hence, both Steiner points and leaf nodes are associated with an upper bound on their maximum degrees on the Steiner tree. The bounded degree minimal Steiner tree (BDSMT) problem is intractable. An ILP-based solution to this variant of the minimal Steiner tree problem has been proposed [17].

Table 3.1 provides the comparison of the wirelength for the routing of two methods for building sensor bias current networks. CMT denotes the current mirror tree topology constructed with the BDSMT formulation [17]. DCS denotes the dedicated current source topology. Three example sensor placements are depicted with the number of sensors varying between two to eight. The sensor network is superimposed on a single core processor floorplan (Alpha EV6) using an optimized sensor placement technique [18]. The goal of the sensor placement algorithm is to minimize the measurement error for thermal profiles under typical workloads with a fixed number of sensors. Using this tool, sensor placements for different number of sensors were generated. Column WL denotes the total wirelength of the current distribution network for each benchmark (the voltage measurement network is not included in the calculation, since it is same in either design). The ILP formulations of the BDSMT instances are solved with CPLEX Ver10.1. The relative saving of wire length can be as much as 42.74%, and is 27.65% on average. Table 3.2 summarizes the statistics of the measurement accuracy when the sensor network was subject to two alternative calibration schemes. The BDSMT infrastructure

Table 3.1 Wirelength for the routing of alternative current distribution networks for thermal sensors. CMT denotes the current mirror tree structure and DCS denotes the topology with a dedicated current mirror for each sensor back-end

Benchmarks		**CMT**	**DCS**	
Name	**#diodes**	**WL (mm)**	**WL (mm)**	**Impr%**
***Rbs*1**	2	12.25	13.59	9.86%
***Rbs*2**	4	21.24	30.49	30.34%
***Rbs*3**	8	40.43	70.61	42.74%

Table 3.2 Examples of representative voltage and frequency ranges used for dynamic voltage and frequency scaling in processors

Processor	**Voltage range**	**Clock frequency range**
Intel Pentium M	0.956–1.484 V	600 MHz–1.6 GHz
Intel Core i7	0.707–1.86 V	1.2–3.2 GHz (increments of 100 MHz) Turbo Boost levels: 3.3 and 3.4 GHz
AMD FX-8120 Bulldozer	0.875 –1.412 V	1.4–3.1 GHz (increments of 100 MHz) Turbo Core can boost up to 4.0 GHz

achieves 1°C average accuracy for the two-point calibration scheme and using the three-point calibration the average accuracy can be less than 0.6°C.

3.1.2 *Communication networks for temperature sensor outputs*

The other critical aspect of the sensor interfaces is the network that conveys the DTS readings to the thermal management module(s). This network will exist for both analog and digital sensors. For digital sensors, each individual sensor may generate a digital output locally, which will then be routed to a control/monitoring module. For analog sensors, either each sensor may possess a local digital conversion front-end or multiple analog sensor back-ends may share a common digital conversion front-end remotely. Another possibility for analog sensors is that the task of digital conversion could be a subcomponent within the thermal management unit.

In the case of analog sensors without a local digital front-end, the transmission of the analog outputs to a remote thermal monitoring and management unit will entail the accuracy concerns described in Section 3.1.1. As we have discussed, adopting the four-wire measurement could help alleviate thermal variation-induced voltage drops across the resistive network carrying the sensor outputs. Also, these interconnects can be sized differently to mitigate additional resistive drop and shielded against noise [11]. For future high-performance chips with increasing number of temperature sensors a hierarchical network with multiple hubs for sensor data collection and conversion might be needed. Furthermore, an existing trend is to deploy a mix of varying sized analog and digital sensors to create a fine grain coverage trading off sensor footprint and routing overheads, and also maintaining a balanced overhead of analog versus digital components.

The interfaces for digital sensor outputs, on the other hand, involve two aspects. First is the physical structure of the interconnect architecture. Second, is the communication protocol for the transmission of data. The physical structure can be broadly categorized in two types: (a) point-to-point communication between sensors and the thermal management unit; (b) utilizing the testing infrastructure or a system bus; (c) a specialized network structure to connect sensors and management units. In the final case, there are emerging instances of network implementations that serve data communication for other types on on-chip monitoring devices (e.g., process monitors, critical path monitors, aging monitors) in addition to temperature sensors [19].

One of the early commercial implementations of DTM in Intel's 90 nm Itanium family of processors used point-to-point connections between the ADC front-ends of four temperature sensors embedded into the floating-point and integer units of each core. This type of sensor interface was only feasible for a small number of sensors deployed in early products. With the increasing number of sensors and elevating sampling speeds expected from on-chip monitors, more structured interfaces have started to emerge.

An early proposal for interfacing on-chip sensors suggested to integrate the temperature sensors with the boundary scan functionality of a chip [20]. By introduction of additional boundary scan instructions the read-out circuitry of the sensors would be enabled or disabled and chip temperature is sampled. Another proposal involved using the joint test action group (JTAG) [21] Test access port (TAP) [22]. The access circuitry is defined using the boundary-scan description language and the JTAG TAP operations are described with the serial vector format [23].

In newer designs, sensors usually include a serial interface, such as I^2C, SMBus, and on-chip peripheral bus (OPB), which provides communication with embedded microcontrollers and other digital systems [13]. The IBM Power6 architecture [2] coordinates thermal sensors, delay sensors, and actuators via the I^2C high-speed serial bus. Addressable registers are used as the interfaces to these components. I^2C bus is commonly used serial data line interface between peripherals and processors with speeds ranging between 100 kbits/s to 3.4 Mbits/s. A dedicated thermal and power management microcontroller communicates with its own fast I^2C bus interface with the sensors. Similarly, I^2C interface is used to establish connectivity with the temperature sensors in the Power7 chips [3]. Power7 implements an additional feature for low overhead communication. An intermediate transaction table periodically collects thermal sensor data from around the chip. Once a certain block of the latest set of sensor data is consolidated in this central location, a dedicated microcontroller in charge of thermal and power management can stream it out as a single I^2C transaction.

Intel developed the SMBus serial interface, which is also compatible with I^2C. Remote analog sensors placed on the die of a high-performance chip can utilize this interface to transfer their readings to a thermal monitoring chip for subsequent conversion and analysis [24, 25].

DTSs instantiated in FPGA chips [26–29] are often comprised of delay line or ring oscillator structures with a time-to-temperature or pulse-to-temperature front-end. They utilize an interface such as the OPB communication channel [30]. OPB cooperates with the processor local bus through a connector bridge to enable embedded processor cores in FPGAs to interface with all other peripherals on the chip. In terms of hardware implementation on the FPGA, OPB specification takes the shape of a distributed multiplexer.

Bus-based communication enables easy integration of sensors into existing communication substrates and compatibility with communication protocols. However, scalability is a major concern with increasing number of on-chip monitors, particularly with the advent of an increasing variety of on-chip monitors continuously reporting status on a multitude of physical events on the chips such as temperature, leakage, delay, voltage drop, current flow. In response to this growing need, specialized NoC structures to collect sensor data have been proposed by recent academic studies.

One of the early proposals for a dedicated NoC for on-chip sensors network involves communication support for two types of on-chip physical monitors: temperature sensors and critical path delay monitors [31, 32]. It is intended to perform monitoring of SoCs via a dedicated processor called monitor executive processor, which performs thermal-aware dynamic frequency scaling on the system. The network uses an irregular mesh topology that conforms to the irregular placement of the monitoring devices on the chip. It is comprised of a combination of multiplexers, bus segments, and routers. Low bandwidth monitoring devices such as temperature sensors are connected to the bus segments via multiplexers, while higher bandwidth sensors are connected to the router ports directly. A low complexity routing protocol has been adopted, where each router is programmed with a static routing table. The routers are optimized for area and their structure is relatively simple in comparison to traditional NoCs for data communication. The routers contain four-entry buffers with 8-bit data width. The number of ports of each individual router is also customized for a given irregular topology.

Another NoC proposed for temperature sensors aimed to optimize the network's complexity by observing that data flow in its main mode of operation will be unidirectional [33]. A tree topology has been used to this end with a central control unit at the root and routers at the intermediate nodes. In order to improve fault tolerance within the network an additional bidirectional link has been inserted between pairs of sibling nodes only. This study explores a range of router configurations and identifies 4-bit data width and a four-entry router buffer as a reasonable design choice. The routers do not employ any LUTs, since data flow (in absence of faults) is unidirectional towards a single destination. Figure 3.6 illustrates the basic structures of these two NoC configurations. Figure 3.6(a) depicts an irregular mesh structure attached to the thermal monitoring unit and Figure 3.6(b) depicts a tree structure.

In emerging high-performance chips, other on-chip monitoring devices such as current and delay sensors are co-located with temperature sensors. A NoC structure was proposed to serve a multitude of such on-chip sensors including temperature sensors [19]. This network introduces the notion of thresholding, where sensor

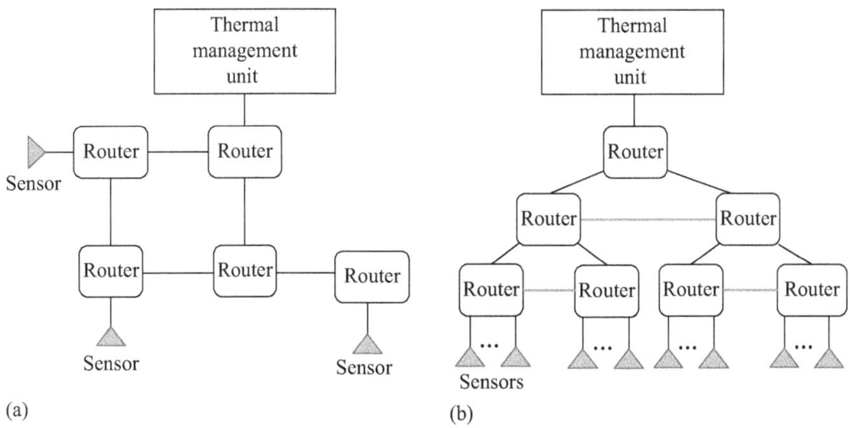

Figure 3.6 Illustration of two alternative NoC architectures proposed for on-chip temperature sensor networks: (a) mesh structure leading to the central thermal monitoring unit, (b) tree structure with special direct links across routers at the same level, sensors attached to the leaves, and the thermal management unit interfaced with the root router

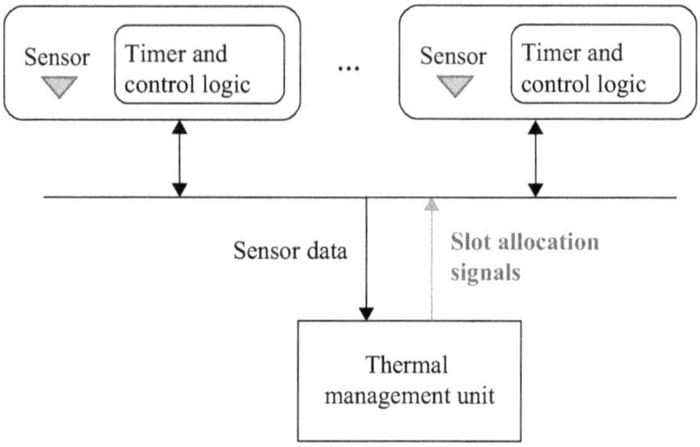

Figure 3.7 Important components and signals communication within the specialized time multiplexed network of temperature sensors

outputs are processed and partially eliminated as they are transmitted through the network and only the most relevant subset arrives at the final destination, that is, the central thermal monitoring unit. Specifically, a monitor threshing level is defined, such that only those sensor outputs exceeding that threshold are allowed to propagate through the network. A shared data channel is time-multiplexed among multiple monitors. Furthermore, unlike the other networks described earlier, the outputs of the sensors are assumed to be analog and they will be digitized at the destination central unit. The network back-end is responsible for allocating a single cycle slot for each monitor and determine the start time of communication with a specific individual monitor via acknowledge signals. Figure 3.7 shows the overall structure of this specialized network carrying analog sensor signals through a shared channel.

All of the network structures reviewed so far are built in a synchronous fashion. A self-timed communication network for sensors has also been proposed [34]. The rationale behind this choice is to enable sensors to immediately alert the control mechanisms of sudden thermal events. While most thermal events will entail a sufficiently large timescale, most likely larger than the sampling period, this network is an interesting concept. This network is a star interconnect network with point-to-point channels from each sensor to the central processing unit. It employs a two-signal handshake protocol between each sensor and a central monitoring unit. Two-bit encoded data dividing the entire operating temperature range into four regions is transmitted from each sensor through four wires by changing the signal on one of the wires to signify one of the four regions.

3.2 Dynamic power and thermal management using on-chip temperature sensor feedback in industrial designs

In this section, we will first review DTM systems with temperature sensors in their feedback loops adopted by industrial designs. Next, examples of academic studies

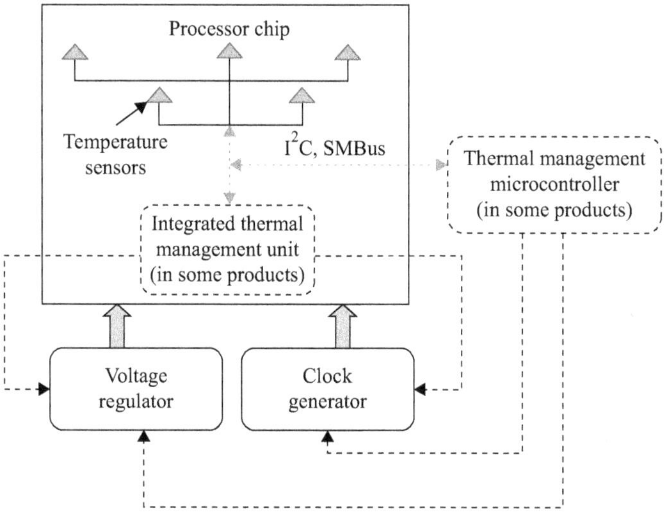

Figure 3.8 A generic illustration of the major actors in DTM

will be discussed. Majority of DTM implementations in commercial chips are fundamentally driven by the die temperature reported from the on-chip temperature sensors. Certain systems also incorporate secondary inputs from on-board sensors. For instance the Intel Xeon Phi co-processor employs temperature sensors at the card-level reporting the inlet air, outlet air, and one placed near the GDDR5 memory chip. Figure 3.6 illustrates major components that are part of the DTM systems in modern processors and their interactions. A specialized module (thermal management unit) is in charge of collecting temperature sensor data and making decisions on management actions. This module might be integrated into processor cores or be implemented as a dedicated microcontroller. Both options are depicted in Figure 3.8.

3.2.1 Early methods for DTM

Early examples of dynamic power and thermal management schemes are those adopted by products such as Intel's Pentium 4, IBM's PowerPC, and AMD's Athlon family of processors [35–38]. Analog temperature sensors built out of thermal diodes are used to provide the necessary feedback for these early control schemes.

Earliest implementations of DTM for Intel's processors involved creation of idle time in response to thermal emergencies to allow for the chip to cool off [35, 39]. The on-chip temperature sensor's measurement is compared against the factory calibrated maximum temperature threshold. This temperature value is referred to as $T_{junction}$ or T_{max} and it represents the maximum allowed temperature for the chip. The actual measured temperature reported by the sensors is represented as a Δ-difference from $T_{junction}$, which can be a vlaue ranging from some positive number to 0. If the Δ-difference is 0 or very close to it this will indicate a trigger for the thermal management system. At the other extreme, when the Δ-difference drops below a certain minimum limit (~20°C below the safe temperature limit) the sensor outputs bottom out until the temperature rises high enough to report an updated new Δ-difference.

The system can be in one of the two operating modes in response to this comparison. If the temperature threshold is not exceeded, the system runs at the maximum clock frequency. If the threshold is exceeded, a signal named PROCHOT# is asserted to activate the thermal control system. In that event, the system stops the clock at a pre-defined duty cycle until the die temperature drops below the emergency threshold and the PROCHOT# signal is de-asserted. Finally, in the event of a cooling failure or other unexpected behavior, if the processor temperature rises above the pre-defined limit of $T_{junction}$ this indicates that the management system is not capable of reducing the die temperature under a safe limit. Actual implementations may vary, but this catastrophic thermal point is approximately at around 20°C above $T_{junction}$. This triggers a so-called THERMTRIP# state and the system shuts down the internal clocks until the processor cools off.

This thermal management mechanism is termed as Thermal Monitor1 (TM1) and it can be considered as the first generation of Intel's DTM systems. It operates automatically under the control of the hardware if enabled. However, it can be enabled or disabled by the OS by setting one of the model-specific registers (MSRs) through the advanced configuration and power interface (ACPI). It can then be enabled at run-time by the OS and this mode also allows configuring the duty cycle of disabling the clock in a wider range. This mode of usage is referred to as an on-demand management. Regardless of the management mode (automatic or on-demand) the duration of the time t_{off} that the clock is disabled is limited to approximately 3 μs. This interval is maintained constant, while the desired duty cycle is achieved by modifying the duration of time, t_{on}, that the clock is active. The resulting duty cycle would be expressed as:

$$d = \frac{t_{on}}{t_{on} + t_{off}} \tag{3.3}$$

A study of the relationship between the clock gating period and the power density in the vicinity of the sensor triggering throttling has observed that power density at the sensor is a square wave of the period $t_{off} + t_{on}$ [40]. The amplitude Δ_T of temperature swing resulting from the change in power density can then be approximated as [40]:

$$\Delta_T(d) \approx \frac{Q_{on} - Q_{off}}{k_{Si}} \times \sqrt{\frac{4\alpha_{Si} t_{off}}{\pi}} \times \sqrt{d} \tag{3.4}$$

The choice for the length of t_{off} involves two considerations. On one hand, the swing in temperature should be just enough to bring the temperature under the safe limit. Yet, it should not be excessive resulting in significant performance degradation. It is generally desirable to operate a system under full computational load, only slightly under the temperature limit, resulting in the highest power and performance setting. On the other hand, the duration should be long enough to allow for the sensors to capture the new thermal state. A few microseconds has been determined to be sufficient for this purpose in practice.

Early IBM PowerPC family of processors used an on-chip TAU that can operate in two modes [41]. The junction temperature reported by an on-chip sensor can be compared either to a single threshold or to two. This enables a choice between a one-step versus two-step intervention. Threshold temperatures are programmed into special purpose registers that are compared against the reported measurement. Upon exceeding a threshold, an interrupt is generated which guides the processor to throttle its instruction cache fetch rate. In some variations of the system the monitoring system has also been implemented as an off-chip monitoring processor, which communicates the thermal diode on the processor chip and send back necessary intervention interrupts through an I^2C bus via an additional bridge chip [42].

3.2.2 Second-generation methods for DTM

These early implementations were extended for newer chip families and also combined with various power management methods. Over the time, manufacturers also started customizing dynamic management techniques towards specific domains such as mobile, desktop, and server. This led way to the second-generation DTM methods that we will review in the following.

3.2.2.1 Intel's SpeedStep technology

Pentium M was an example for achieving better efficiency in the mobile domain [37]. The second-generation DTM scheme, termed Thermal Monitor2 (TM2), was introduced starting with this processor family. Unlike TM1, which only disabled clock in short intervals according to a duty cycle schedule, TM2 implements dynamic voltage and frequency scaling in response to thermal events and/or utilization. Similar to TM1, there are two modes: hardware-based automatic mode and the on-demand mode controlled by the software.

This processor chip employed two sensors of each type: one thermal diode and one DTS. The thermal diode operates in remote mode, that is, only the back-end resides on-chip. Its output is relayed to an external chip for A/D conversion via dedicated pins. This sensor is tasked with reporting the junction temperature to the OS. The second sensor is a DTS that is interfaced with the on-chip management mechanism.

When the DTS signals that the chip temperature exceeded the threshold, the thermal management mechanism is engaged for a time period (at least for about 1 ms). The clock frequency is changed first. The phase locked loop (PLL) circuitry is locked onto the new frequency level. This transition happens within a few micro seconds, hence the execution of the processor is not interrupted. Once the frequency is stabilized, the voltage is reduced to the corresponding frequency at an approximate rate of 1 mV/μs. There may be variations to the specific implementation. One option has been to maintain a list of (frequency, voltage) pairs and move from one to the next over time in small increments. Each (frequency, voltage) pair is also referred to as a *P-State*. Intel's power and thermal management system defines several P-Sates (~5 states) and the highest performance state is named P0 and P1 is the next level of performance state and so on. These steps are usually taken one at a time dropping the frequency and voltage gradually by the hardware. There is also a sequence of states, called the C-States, which correspond to different configurations of idle states. C-States vary in the specific subcomponents that are placed in sleep mode (e.g., only caches in sleep mode versus core and caches).

During the time period allocated for thermal management, the DTS continuously reports temperature and the management system can infer whether its measures have successfully reduced the temperature below the threshold. If this is achieved soon enough, the thermal management mechanism can step aside before the end of the 1 ms period. If, on the other hand, the chip temperature continues to rise despite operating at the lowest (frequency, voltage) point due to a malfunction or cooling problem, the processor is placed in shutdown mode as in the TM1 method. In a symmetric fashion, the frequency and voltage of the processor can be ramped up during periods of a large gap between the current and maximum junction temperature to optimize performance.

The dynamic power and thermal management scheme combining the first and second generation thermal monitoring methods TM1 and TM2 is also referred to as the SpeedStep technology. The hardware-based self-monitoring and self-throttling is complemented with software-level optimizations. While the on-chip monitor is tracking the digital sensor, the output of the thermal diode is being translated by an external A/D conversion chip and its digital output can be obtained through the ACPI. The OS utilizes this junction temperature information to implement an on-demand power optimization scheme. This is also referred to as the demand-based switching (DBS). DBS aims to step down the P-State of the processor when it recognizes an opportunity for low utilization. At times of increased workload it elevates the P-State to match the demand. At the software layer, more sophisticated custom voltage and frequency scaling algorithms can also be implemented to manage the power and thermal state of the chip at a coarser level of granularity. The hardware-based monitor is then largely in charge of responding to sudden changes and malfunction rapidly ahead of the software layer.

3.2.2.2 Intel's Foxton technology

Another early example to specialized thermal management is Intel's Foxton technology [43, 44]. Ninety nanometer dual-core processor families employed a dedicated embedded microcontroller interfaced with the on-chip analog sensors. The monitoring system is a discrete time control system. It can operate in a fixed mode, where the voltage and frequency is fixed without any dynamic management. It can also operate in the dynamic management mode, where the microcontroller's firmware is capable of executing three different routines: power, temperature, and calibration. These routines are invoked via a real-time scheduler at pre-defined intervals (8 μs). This system employed four temperature sensors located in floating point and integer units. These sensors are comprised of forward-biased diode junctions and constant current sources, with an approximate temperature coefficient of −1.7 mV/°C. The analog sensors are interfaced via multiplexers with ADCs located near the Foxton unit. The ADCs are shared with voltage measurement unit and other management and calibration-related signals, which the microcontroller also needs to take into account. The microcontroller monitors current consumption of the package. Voltage measurements across a special power plane resistor are utilized to derive current and subsequently power consumption of the package.

It has been reported that the routing from the remote analog sensor back-ends to the centralized ADC front-ends required careful planning to avoid resistive voltage drop along the lines and noise coupling. This points to the importance of the routing network as we have also discussed earlier in this chapter.

The power and temperature management system operates with the objective of maximizing performance while remaining within defined temperature and power bounds. This processor family had a power consumption ranging approximately between 100 and 130 W. This system is configured to not allow the chip temperature to exceed 90°C. During each power routine, the current measurements are converted to a power measurement by the monitoring system. A closed loop control system determines the voltage level for the package for the next epoch such that the system would stay within a defined power envelope defined by a target reference power. The system can utilize a 32-step voltage scale with 12.5 mV resolution. The resulting voltage levels are then passed onto a subsystem, which collaborates with the clock generation and is in charge of determining the required clock frequency for that voltage level [45]. There are 64 available frequency levels. The temperature routine collects the actual sensor readings from the ADCs and compares them against the threshold. If the management system decides to intervene for a thermal emergency, it accomplishes this by placing a cap on the highest voltage level that the power routine can work with. Finally, the calibration routine is intended to calibrate voltage measurements only and does not involve sensor calibration. While the Foxton technology can be used purely in hardware-driven mode as described above, it also allows OS access and control to execute in DBS mode.

3.2.2.3 IBM's EnergyScale technology

IBM introduced the EnergyScale technology for dynamic power and thermal management for its dual-core Power6 processors [2]. This system, similar to Foxton, utilizes a special microcontroller named the thermal and power management device (TPMD). Performing on-chip thermal management as opposed to a purely software-based control enables rapid run-time response to thermal emergencies.

Two types of temperature sensors are available on this processor and they communicate with the TPMD via the I^2C bus. The first set is comprised of analog on-chip thermal sensors implemented as a thermistor. The sensor is calibrated as a resistance with linear dependency on temperature. The current through the resistor subject to a constant known voltage across its terminals is measured and converted to temperature reading. On the dual-core chip, there is one thermistor placed on each core and a third thermistor is placed near the center of the chip. The processor contains dedicated pins for these sensors to route their current measurement to an off-chip ADC. Power6 is one of the modern examples of a processor chip with enhanced and fine-granularity on-chip temperature sensing. In order to achieve this, a total of 24 DTSs (eight in each core and eight in the remaining chip area) have been placed on the chip. These sensors are directly accessible on-chip for the hardware-based thermal management system. The digital sensors have been implemented as ring oscillators, where the temperature dependence of the oscillator's speed is captured with the count of oscillations over a fixed period of time. Since this chip employs dynamic voltage and frequency scaling, the supply levels are obviously expected to fluctuate significantly. The ring oscillator design has been therefore specially optimized to be voltage invariant.

The digital sensors are interfaced with a specialized network that transmits the oscillation counts reported by them into special system registers. Thereby, a dual-mode operation is achieved similar to the aforementioned examples of other processor families. The on-chip thermal monitoring mechanism provides its feedback to the TPMD and receives directives on how to remedy the thermal emergency from the power and thermal management firmware of the microcontroller. In order to accomplish this task, TPMD executes a real-time embedded OS, polling sensor information (both analog and digital sensors) and making management decisions. The processor's OS also has access to the status data of the TPMD.

The TPMD executes a so-called thermal capping scheme. There are two options available for thermal management specifically: thread-level pipeline throttling and frequency scaling. These options can be used in isolation or in combination. When the TPMD detects that the chip temperature has reached a critical level it directs the hardware to reduce the dispatch rate of the threads running on the cores. The dispatch rate can be elevated back to the active rate after the sensors report a sufficient reduction in temperature below safe levels. The second option for intervention is throttling of the clock frequency and configuring the voltage setting accordingly.

3.2.3 Latest generation methods for DTM

A trademark of the latest generation dynamic thermal monitoring methods in current practice is the widespread use of all digital on-chip (i.e., both front-ends and back-ends implemented on-chip) temperature sensors. For instance, IBM's 8-core Power7 processor utilized 44 on-chip digital sensors [3, 4]. We also observe that factory calibration of each individual on-chip sensor now occurs by default and this is largely driven by the stronger die-to-die variations at scaled processing nodes at which these processors are built. Another important feature of the latest generations of processors is the use of the Turbo Boost mode [46], which introduces an additional frequency level above the nominal to boost up performance opportunistically whenever the thermal and power envelope of the system provides sufficient slack. For instance, AMD processors utilize two new features named the HY Boost and Configurable TDP. HY Boost utilizes on-chip temperature feedback when considering the timeliness of initiating a Turbo Boost phase. The thermal state provides the most accurate status of the underlying chip as to the available thermal headroom compatible with the target TDP envelope.

Finally, as more of the latest processors move from passive heatsink-only cooling to active fan-based cooling we start to observe the impact of the on-chip thermal monitoring on the cooling system. Control of the active cooling also becomes part of the on-chip thermal monitoring systems' jurisdiction.

3.2.3.1 IBM Power7 and Power8 thermal management

IBM deployed an enhanced version of their EnergyScale technology in the newer processors utilizing a dedicated microcontroller (named the EnergyScale microcontroller) for power and thermal management [3, 4]. A total of 44 DTSs have been distributed to 8 cores (five per core) and the periphery (remaining four sensors). One notable difference is the additional effort placed into calibrating these sensors with a more thorough process. The digital reading of each sensor is further processed by the EnergyScale microcontroller with a quadratic polynomial to map it to the ultimate temperature value in °C. This polynomial is configured in the microcontrollers firmware through individual calibration of each digital sensor in the factory against a reference thermal diode and within the expected high range of operation (65°C–85°C).

The power management decisions to respond to thermal emergencies are then applied based on the feedback from the sensors (e.g., dynamic frequency and voltage scaling, activity throttling). A distinct advancement in these policies in comparison to the previous processor generation is the ability to scale the frequency of each individual core separately. Power7 provides a dedicated frequency generator for each core utilizing a fractional digital PLL. Each generator uses a digitally controlled oscillator to create a number of multiples of a reference clock. The latest member of the family, Power8, a 12-core processor fabricated at 22 nm process node, also uses an on-chip controller to execute similar power and thermal management services [47]. The most distinctive feature of this latest generation dynamic management system is that the control loop executes in under 1 ms, while it took about 32 ms for Power7 system.

3.2.3.2 Intel Xeon and core families

The 45 nm Intel Xeon processors [48, 49] and 22 nm Intel core-family [50] of processors utilize one on-chip DTS per core that are factory calibrated. Similar to previous generations, these sensors report a relative die temperature as an offset from a programmed maximum temperature threshold $T_{CONTROL}$. Under no thermal stress the sensors report a negative reading and the thermal management is triggered when the difference reaches zero. One modification in this system is that the temperature limit to trigger the PROCHOT# state and thermal throttling has been elevated compared to previous designs. This helps to absorb variability in the absolute trigger limits experienced across different processors and provide a more uniform thermal management experience to the end users. For instance, an important concern is the acoustic implications of the cooling fans. As will be discussed shortly, the temperature readings of the on-chip sensors are also utilized to control fan speed by the system software. Large variability in the fan speed control across different processors would result in drastically different acoustics outcome.

A modification to the PROCHOT# signal has been made in comparison to the Pentium M family to better suit the needs for managing multiple cores. In the new implementation it is realized as a bidirectional signal. When a thermal emergency is triggered by an on-chip sensor of a particular core, this signal is asserted by the processor and any performance throttling to tend to the thermal emergency would take place locally within the corresponding core. Assertion of the PROCHOT# signal in the opposite direction by an external entity would invoke throttling across the entire chip.

As far as the specific thermal management schemes are concerned, the sensors are interfaced with an on-chip embedded thermal management circuitry, along with the sensors they are collectively referred to as the DTS circuitry. They are comprised of a micro-controller circuit and clock controller. This autonomous hardware unit can execute one of the two management schemes, TM1 and TM2, previously introduced with the Pentium 4 and Pentium M families, respectively. Also, similar to older generations, the THERMTRIP# signal is used to indicate that the processor's junction temperature has reached a point beyond the reliability guarantees, possibly due to a localized thermal runaway or cooling failure. In response to this event the shutdown process will start, first with the internal clocks being stopped and next the core's voltage regulators being disabled.

At the software-level, basic input output system (BIOS) can enable or disable the TM1 and TM2 schemes. For system-level management, the platform environment control interface (PECI) is used to communicate the thermal state to the software. PECI is an Intel proprietary interface that provides a communication channel between processors and chipset components to external thermal monitoring devices. PECI essentially acts as a bridge between the software layer and processors. Digital sensor outputs are stored in special MSR registers like in previous generations. These are interfaced with PECI, which applies additional processing on them to compute the average temperature of the hottest core. While PECI has access to all digital sensor outputs, software only has access to this single register representing the maximum temperature. The individual MSRs can be read through special instructions as well by the software for advanced custom power and thermal management schemes. There are also various third party utilities to read these values, such as the Windows-based Real Temp and the Linux Kernel Module Core Temp.

The physical structure of PECI has been modified to improve the scalability of the system with increasing number of cores and corresponding number of on-chip thermal sensors. The network of sensors is divided into domains in a hierarchical fashion. For example, each socket can constitute one domain, or a socket can also be divided into domains internally. Each domain is represented with the maximum temperature value reported from the sensor of the hottest core in that domain. Each domain can now be polled individually through the PECI. The PECI output is also used to control the cooling fan speed. While PECI reports a temperature below $T_{CONTROL}$ the fan speed can be reduced for improved acoustics and power savings. Upon triggering of the PROCHOT# signal the system fans are ramped up to full speed.

3.2.3.3 Thermal management on the Intel Xeon Phi coprocessor

Our discussion on DTM mechanisms for commercial processors will be concluded with the final example of the Intel Xeon Phi coprocessor as an example of first generation many-core chips and the associated card design [51]. This system illustrates the integration of on-chip thermal sensors with additional on-board thermal sensors for system-level management. The main control knob for responding to elevated chip temperatures is frequency scaling in this system. The coprocessor chip contains 61 cores, each with a factory calibrated DTS. The on-chip DTS circuitry similar to the one described for the Xeon and Core families of processors is also deployed on these coprocessor chips. The PROCHOT# and THERMTIP# signals function in the same way as described earlier to trigger throttling or shutdown. The first stage of throttling is a stepwise reduction of processor frequency in increments of 100 MHz for a pre-determined amount of time. If the reported temperature still does not drop below $T_{CONTROL}$, then the OS complements this with a reduction of voltage. The rolling average of the highest reported sensor reading from the many-core chip is transmitted to an on-board System Management Controller chip, which also receives the data from the other on-board temperature sensors as well as voltage and current monitors through the I^2C bus. It is also interfaced through the SMBus with the system software to report thermal state and receive management directives from the OS.

Table 3.2 depicts voltage and frequency ranges used for three commercial processors; an older processor (Pentium M) in comparison to two newer chips (Intel's Core i7 and AMD's Bulldozer). Power and frequency settings of cores in these processors are managed by their respective mechanisms within these depicted ranges.

3.3 Methods on the use of on-chip temperature sensor feedback for dynamic optimizations beyond commercial designs

Existing commercial implementations of DTM with on-chip temperature sensors in the feedback loop focus exclusively on tracking the single hottest temperature representing the overall thermal state of a chip. The response mechanisms (performance throttling, dynamic voltage, and frequency scaling) are also core-wide or chip-wide. Over the last decade and a half DTM has developed into an active field with a diverse set of proposals to mitigate the thermal control of high-performance systems [52]. These emerging proposals mainly formulated a variety of micro-architectural or system-level ideas to complement the coarse-grain voltage and frequency scaling methods. In the following, we will discuss prominent examples of these efforts from academic research.

3.3.1 Hardware-based thermal management

DTM involving hardware-based throttling in the academic domain involves various enhancements to the dynamic voltage and frequency scaling and clock throttling methods and more importantly they made significant novel contributions to the related domain of microarchitectural techniques. The latter ones introduce dynamic management of major architectural blocks (e.g., caches, fetch, decode, and speculation units) in response to temperature sensor feedback. These works can be further divided into two major subcategories:

1. Proposals that perform additional processing (such as filtering, extrapolation) of physical temperature sensor data;
2. Proposals that aim to create correlation between higher level activity indicators and temperature and still use physical temperature sensor feedback as a means to train the models representing such correlation.

In the following we will review major efforts that involve hardware and architectural features for DTM with on-chip temperature sensor feedback from these three main points of views.

3.3.1.1 Involving architectural features in thermal management

An early study on a number of potential approaches to DTM argued for the need of quantification of the correlation between the traditional throttling schemes and their impact on performance [53]. Leveraging such established correlations, alternative management methods were proposed. In addition to the existing methods of voltage and frequency scaling (applied in majority of commercial products), instruction cache throttling (PowerPC), and frequency scaling (applied in some commercial products), two new microarchitectural solutions were proposed. The first method restricts the speculation window of the execution when the thermal sensor feedback indicates that a safe limit has been exceeded. The other method is called the Instruction Cache Toggling, which disables the instruction fetch unit and the branch prediction unit for a specific number of cycles. In that period, the processor is fed from the instruction fetch queue. In this study, emphasis was placed on the role of the response delay of the DTM mechanism in the effectiveness of the thermal management. This concept is illustrated in Figure 3.9. Two thermal response curves are depicted, with and without DTM. Without DTM, the system

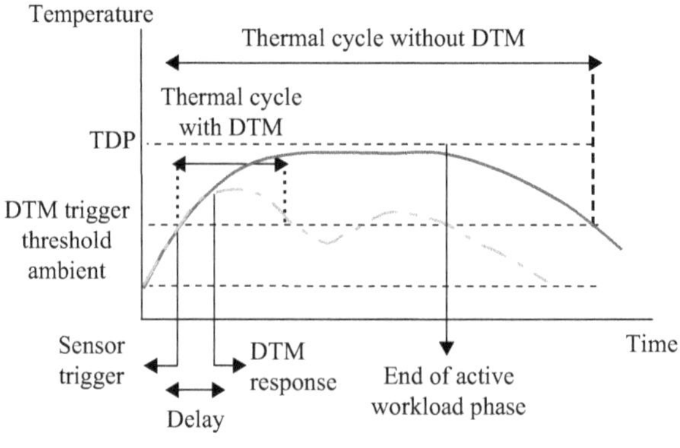

Figure 3.9 The impact of DTM's response delay on the thermal state of a system

(if the package and cooling is designed properly and TDP is calculated carefully) reaches the thermal design point and stays at this course until the workload's active phase concludes with a low activity or idle period. If the system is using DTM, the sensors will notify that the DTM threshold has been exceeded, yet the response will activate with a delay. Due to this delay, the system will spend a period of time above the safe limit, which can be associated with the duration of a thermal cycle experienced by the system. Based on this observation, a solution was proposed to minimize the overhead of invoking the proposed microarchitectural management schemes. Specifically, it has been proposed to feed the on-chip sensor data directly into the state machines of the corresponding components.

This work is significant in that it is one of the first attempts to systematically quantify the impact (reduction in performance) and benefit (effective number of cycles spent in thermal emergency state over all execution cycles) of different management schemes. The main finding is that the thermal benefit of the management scheme is tightly coupled with its activation latency. In that regard, voltage and frequency scaling seemed to be at a disadvantage. On the other hand, finer grained microarchitectural methods provide higher benefit per unit performance loss. While these general observations still hold true, there has been significant advances in circuit design for voltage regulators and clock generators, which helped to reduce the negative impact of the initiation latency for scaling the voltage and frequency. A related work specifically incorporated the non-negligible transition delay of voltage and frequency scaling into the optimization and proposed optimal selection of a frequency level for throttling among a set of discrete values under a realistic transition overhead model [54]. The main result of this study suggests operating a microprocessor at high speed until the temperature nears the maximum allowable temperature. Once the sensors trigger a thermal limit, the thermal management simply fluctuates between two voltage/frequency levels; the high and low levels over the entire remaining execution time.

Another pioneering work focused on the goal of co-optimizing energy efficiency and thermal stress [55]. In broad terms, this study creates a prioritized list of a large number of possible intervention methods and applies them individually or in groups, selecting the proper priority method according to the severity of the thermal emergency. Techniques are added to or removed from a working set as the temperature reported by on-chip sensors rise above or fall below certain limits. In the case of a fully populated working set activating all possible methods still fail to reduce the peak temperature, then, the system relies on a failsafe mechanism such as the THERMTRIP# state of commercial processors. This study adds a few new techniques to a set of techniques to evaluate their algorithm. Their new techniques mainly focus on the dynamic management of the power of the memory hierarchy such as cache subbanking and lowering the supply of the DRAM.

A spatially aware performance throttling method introduced the concept of physical awareness during frequency selection for DTM [56]. This method provides a systematic approach that can help determine the best operating frequency level for the core in thermal emergency during performance throttling. This entails to bring this core out of thermal emergency within a given time period while still maintaining its operating frequency as high as possible. Thereby, the impact of throttling on its performance would be minimal. This frequency selection method presents two important capabilities. First, it considers physical characteristics of the system, that is, the thermal interaction between physically adjacent cores during thermal emergency management. Second, it offers a fast technique for an optimization scheme to minimize the negative impact of frequency throttling onto the performance of the core that is in thermal emergency.

Another spatial approach introduced an architectural tool for performing activity throttling without first resorting to a chip-wide voltage and frequency scaling. This method has been termed as the occasional-2-cycle operation [57]. It focuses on the spatial aspect of thermal events and re-designs functional units (adder and multiplier) so that they can be supplied selectively with a scaled voltage, while still maintaining the same chip-wide frequency. Majority of thermal peaks are expected to occur in the vicinity of these units. While hotspots can also arise near instruction fetch units and register files, this is a reasonable approximate assumption and can mitigate at least a subset of thermal events without requiring chip-wide throttling. When operating at a scaled voltage, the slack in off-critical paths is expected to keep the latency of the units in a single cycle, while only occasionally they would require two cycles (if one of the few critical paths are activated). These units are controlled with temperature sensor feedback and when the sensors trigger indicating that the thermal limit is exceeded, then, these functional units switch to a low voltage.

Finally, there is a large body of work focusing on proactive thermal management relying purely on performance counters and other related system events and software-based power and temperature estimations. Since our scope is focused on temperature sensor feedback and its role in thermal management of circuits and systems, we shall acknowledge this other area of research, but will not review proposals in this domain in detail. However, there are works that combine offline prediction/pre-optimization with run-time on-chip temperature sensor feedback as well. One example is the two-phase DTM scheme [58]. During the first offline phase optimal frequency settings of each processor core is calculated, if execution had started from a certain temperature such that the steady-state temperature reached would remain below a safe threshold. This information is stored in a table for a number of frequencies and start temperatures. During the second phase at run-time, the on-chip temperature sensor for each core is sampled and the table from the first phase is accessed to lookup the best frequency that still keeps the core under the thermal emergency limit for the next interval. Another interesting indirect approach to performance throttling involves a single sensor's feedback getting converted to power, which in turn is re-mapped to a multi-grid thermal map by the system software [59]. This thermal map is then used to perform throughput management of a thermally constrained multicore processor.

The reliability aspect of on-chip heat accumulation led to a revisit of performance throttling mechanisms. One approach to incorporating reliability is based on the observation that periods of intense thermal stress can be compensated by periods of low thermal profiles. A carefully tuned reliability model (mainly representing the impact of temperature in electromigration) is deployed in the management loop, which uses current temperature status of a chip and guides the dynamic voltage and frequency scaling decisions accordingly [60]. The reliability model helps to define a so-called lifetime bank deposit, which is depleted as the chip spend more time under higher temperatures and replenished as the chip spends more time at lower temperatures.

It is interesting to note that performance-centric arguments for intentionally pushing chips beyond their safe thermal limits for brief periods of time have also been made. Turbo Boost technology in newer generation processors is also likely to cause such a thermal response. One example to this point of view is the concept of soft thermal limits [61]. For a given degree of flexibility in the thermal limit during a specific length of time, the proposal method provides an optimization problem and its efficient solution to maximize the clock frequency allocated during this period. We note that, this method might produce even more interesting results when compared with a reliability model such as the aforementioned lifetime banking.

With the introduction of per-core voltage regulation and clock generation capabilities in newer processor generations, attention has been directed towards exploiting this feature for a finer grain control over DTM. In this context, performance throttling can also be considered as a multi-scale and distributed problem. There are design choices in terms of the granularity, for example, per-core versus multiple cores. Quantification of this multi-scale solution space coupled with a sensor-based process migration policy has been investigated [62]. The OS-led migration scheme is established as an outer loop, where the inner loop is a control theoretic dynamic voltage and frequency scaling scheme (applied either per-core or globally). Both the inner and the outer loop can be triggered by a sensor. Another study arguing for the benefits of distributed throttling of cores for thermal management proposed a feedback controller design that uses temperature sensor outputs and avoids the thermal limit being triggered while maintaining a performance goal [63]. Distributed DTM has also been recognized as being especially suited to mitigate unfair treatment of simultaneously running threads in simultaneous multithreaded (SMT) architectures, where an intensive thread causing a thermal trigger can unfairly penalize the performance of other threads that are not likely to cause a thermal emergency on their own. The clustered configuration of SMTs has been exploited to identify hot clusters from their respective temperature sensor feedback [64]. If violation of the thermal limit is indicated by a cluster's temperature reading, then, a distributed technique only throttles the dispatch rate of that particular cluster locally, while the other colder clusters and their corresponding threads can progress.

Another technological development created a new avenue for DTM. Emerging 3D IC technology has brought about new CMP concepts, where a number of cores and a large capacity of memory can be integrated for a tightly coupled system with high data communication bandwidth. Given the additional stress caused by layering several high activity tiers with a weaker thermal conductivity to the ambient along the vertical axis, thermal management for 3D chips attracted significant interest. One of the early studies proposed both OS and specialized hardware support for 3D

CMP thermal management [65]. The resulting DTM system controls performance throttling globally and workload assignment locally.

New circuit implementations for thermal monitoring hardware to support the abovementioned management schemes have also been developed in the course of time. An integrated thermal monitoring hardware comprised of several components in charge of sensor data acquisition, programmable settings for thermal limits, comparison of sensor outputs to programmable registers, thermal emergency interrupt generation, and fan control has been proposed [66]. This work is unique in its consideration of cooling fan control with a hardware-based mechanism, while prior implementations were broadcasting temperature data to the OS and delegate the fan control task to the software layer.

For a specific application, domain-specific methods can also be adopted. One example is the modified decoding sequence for moving picture experts group (MPEG) decoding application [67]. The application's intensive activity falls during the periods of decoding a frame. This time period is being compared against the expected rise time of temperature for the corresponding activity with an analysis. The period of decoding activity is then divided into shorter sub-periods interleaved by idle time, such that the new decoding intervals are shorter than the time by which the temperature is likely to rise up to a level for sensors to trigger thermal throttling. This will inevitably lengthen the total amount of time spend for decoding a frame. Since the system is subject to deadlines for completing decoding, some frames might miss their deadlines. Those frames are dropped by the application. The final analysis shows that this throttling scheme can ensure safe temperature limits with minimal triggering by sensors at a reasonable degradation of output quality and performance.

While direct temperature measurements from well-designed and strategically placed temperature sensors are the most reliable representation of the thermal state, the reliability of sensor readings may exhibit certain variation across different chip implementations. This translates into uncertainty in the temperature observation. A stochastic decision-making mechanism based on a semi-Markov decision process has been proposed to perform the throttling actions [68]. At pre-defined decision epochs, the thermal manager receives temperature sensor feedback. The sensor reading is translated probabilistically into one of three ranges. Each such temperature range corresponds to one state of the thermal manager. The manager can choose the performance throttling action (or any additional management actions) to respond to this observation. Each action is tied with a probability of bringing the manager into a different state.

3.3.1.2 Using post-processed on-chip temperature sensor data for thermal management

The coverage of on-chip temperature sensors has improved over the last decade with the advances of area-efficient and high-accuracy sensor design techniques. Nevertheless, it is not realistic to expect an extremely fine-grain thermal map from existing solutions. There have been many proposals to optimize the placement of sensors and ways to explore alternative positioning of sensors either non-uniformly or as a uniform grid. These options were discussed in Chapter 2. Now, we turn our attention to computational techniques, which aim at improving the ultimate measurement and tracking accuracy from a given sensor deployment. In most practical cases this corresponds to accurately measuring the hottest temperature on the chip at any given point in time, since this value is generally used to trigger a thermal management event. Efforts towards extracting the most detailed thermal map from a coarse grain sensor grid usually are coupled with a thermal management scheme that they support. Hence, to a large degree they can be considered as complimentary to all DTM techniques with temperature sensor feedback. One point to note is that most of the refinement methods rely on sophisticated optimization methods involving floating point computation. None of the proposals to date have provided a dedicated hardware implementation to carry out the necessary computation due to potentially high hardware cost. Hence, these methods require software implementation. This implies that they are likely to be utilized as a guiding input to software-level DTM mechanisms at a larger timescale. On-chip hardware-based thermal monitoring circuitry that serve as the first and most immediate response to thermal emergencies would be unlikely to be driven by the information gathered from the refinement methods due to the mismatch in timescales. On the other hand, experimental evaluations of these methods have demonstrated that these methods can be efficiently implemented in software to perform under the OS's real-time constraints. Hence, at run-time, sensor data refinement methods would collect input from on-chip temperature and power sensors (or power estimators) to perform the translation of the reported sensor data from a limited set of sensors into a wider scale thermal map. The software-level DTM can then utilize this information to perform their management scheme of choice. Overall, the methods that will be described in the following are promising in improving the software-based on-demand thermal management.

A popular method for deriving accurate and fine granularity temperature information from a limited number of temperature sensors is the use of Kalman filtering [69] due to its relative simplicity in comparison to more complex nonlinear estimators, which renders it highly suitable for real-time use, such as post processing temperature sensor readings for DTM. Kalman filter is a well-known tool for modeling systems that are represented with noisy observation data. Kalman filter essentially provides a good estimator of the data observed and measured under uncertain and noisy circumstances. This is viewed as a form of filtering out noise and hence, the method is referred to as a filter. It enables to create a model, which continuously updates its view of the underlying system by evaluating the difference between predictions and the next observed state. Different proposals involving this computational method were proposed [70–75]. An early study of this method adopted the two-step procedure [70]. The first step is performed offline, where an RC-equivalent thermal network of the system is created. The network is represented as a state space with each node's power as its input. A state-space model is generally used to replace an nth order differential equation with a single first-order matrix differential equation. In this case, the temperature at the nodes is the output of the state-space model, where some of the nodes' temperatures (assuming the grid of the model will be finer than physical sensor availability) will also correspond to observations received from temperature sensors. This basic model is in the form of two equations, one representing the time-dependent relationship between the state matrix, input matrix, and the state vector and the other representing the relation between the state vector, input matrix, a transition matrix, and the final state [70]. The first relationship takes the form of a continuous time first-order differential equation as shown in (3.5) in this specific problem.

$$\frac{dT}{dt} = -\left(C^{-1}G\right)T(t) + CP(t) \tag{3.5}$$

In state-space model terminology, the vector of node temperatures $\boldsymbol{T(t)}$, is the time-dependent state vector, the term combining the inverse of the thermal capacitance matrix $\boldsymbol{C}$ and the thermal conductance matrix $\boldsymbol{G}$, is the state matrix, which is a constant, and $\boldsymbol{P(t)}$ is the time-dependent vector of power consumption at each node, which maps to the input matrix of a state-space model.

The second relationship transforms the input vector into a state output, which corresponds to the temperature values at the sensor locations, as shown in (3.6).

$$S(t) = \boldsymbol{F}T(t) + \boldsymbol{D}P(t) \tag{3.6}$$

$\boldsymbol{F}$ is the constant output matrix and $\boldsymbol{D}$ is the so-called direct transition matrix, which is also constant. Systems that do not possess a feedthrough path for the input to directly affect the system output the $\boldsymbol{D}$ matrix can be assumed to be the zero matrix. This assumption has been utilized by this particular approach, which eliminates the second term of (3.6) in practice. The resulting continuous time model is first discretized by replacing the time variable t with a time step variable n and replacing the time derivative of the temperature (i.e., state vector) with the discrete estimation of T at step $n + 1$. Furthermore, an additional term to represent inaccuracies in modeling the thermal equivalent network ($\phi[n]$) and sensor measurement noise ($\sigma[n]$) in the system are added to each equations. Note that it is assumed that the inaccuracy in the model will mainly manifest itself in the DC component corresponding to the conductance matrix and the transient component represented by the capacitance matrix can be ignored. Now, the model takes the following shape, shown in (3.7) and (3.8).

$$T[n+1] = -\left(C^{-1}G\right)T[n] + CP[n] + G\phi[n] \tag{3.7}$$

$$S[n] = FT[n] + \sigma[n] \tag{3.8}$$

Kalman filtering requires two additional pieces of information to perform the estimation. The measurement noise covariance and modeling error covariance are also needed to complete the formulation. With these two known covariance matrices and the discretized state-space representation Kalman filtering aims to minimize the covariance of the estimation error. The actual estimation is performed in two phases. At each time step n, the estimation of temperature from temperature values up to step $[n - 1]$ is updated by the current observation from sensors at step

$[n]$. This updated value of $T[n]$ is then used to estimate $T[n+1]$. This constitutes the first phase, called the time update phase. In the second phase called the measurement update phase, the new sensor measurement $S[n+1]$ is used to adjust estimation of $T[n+1]$.

In an attempt to reduce the computational complexity of this method a simplification has been proposed with the assumption that the sensor noise and modeling error covariances are time invariant. Thereby, they need not be computed at run-time. This improves the complexity of Kalman filtering from a cubical polynomial to a quadratic relationship with the size of the model. This in turn necessitates a one-time calibration to establish a steady-state filter. At the same time the capacitance and conductance matrices of the equivalent thermal network is simplified through model order reduction. In this specific method a projection-based implicit moment matching technique [76] has been used.

Another related proposal aimed at mitigating uncertainties of temperature measurements and probabilistic behavior of power consumption through an intermediate layer of temperature estimators based on a combination of Kalman filtering and Markovian decision process model [75]. The Kalman filter is utilized in a similar manner for the same purpose of estimating temperature based on feedback from noisy sensors as the previously described estimation method. The Markovian decision process is used in addition to probabilistically predict the power consumption of the system under monitoring. The estimations collected from both components are combined to predict the formation of a hotspot in the chip.

Follow up studies have further advanced the body of knowledge on the application of Kalman filters as well as other statistical methods on refining on-chip temperature sensor feedback [74, 77–80]. One of the major enhancements concerns the incorporation of leakage power into the temperature estimation. In the case of thermal monitoring, it is argued that the linear Kalman filter model would underestimate temperature, since it does not consider the leakage power consumption in its input matrix of the state-space model. Since leakage has a non-linear relationship with temperature, the previously developed linear Kalman filter estimators would underperform. With this motivation, a Taylor-series expansion to factor in the first-order contribution of leakage has been proposed resulting in an extended Kalman filter estimator [73]. Recently, a method for sensor feedback refinement specifically for stacked chip technology targeting a 3D multi-processor system of a chip (MPSoC) has been introduced. This method also aims to incorporate the impact of leakage into the estimation using the unscented Kalman filtering method [74]. Unscented Kalman filtering has also been used for online temperature prediction for a planar 2D chip implementation of a processor with the similar goal of accurate representation of leakage related non-linear behavior [81]. This approach could be considered more accurate overall, since the unscented Kalman filtering deals with probabilistic behavior, which undergoes a non-linear transformation. Thereby, the non-linear behavior is maintained as intact rather than being approximated with a first-order expansion. The design methodology for building Kalman filters is summarized in Figure 3.10.

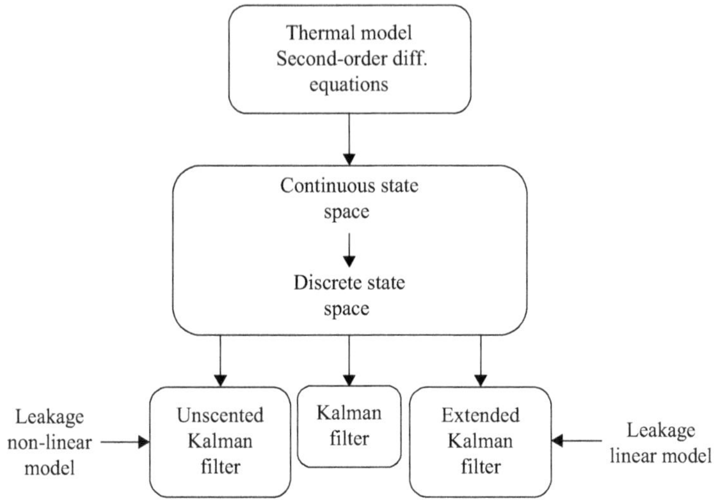

Figure 3.10 Methodology for constructing various types of Kalman filters for post processing of temperature sensor data

3.3.1.3 Building and training models to correlate system events with thermal state for DTM

Soon after the semiconductor industry and research community realized the increasing importance of thermal monitoring and management, a parallel thread of activity sprung revolving around software-based temperature prediction. We do not discuss this aspect in detail since it does not relate directly to the role of on-chip temperature sensing.

Especially in the early years of development of on-chip sensor technology, direct sensor feedback was not as reliable. This was due to various factors including the small number of sensors deployed in early products and the lack of sophistication in early temperature sensors in terms of area and power efficiency and robustness against process, voltage, and thermal fluctuations. Over the last decade temperature sensor design techniques have advanced tremendously and manufacturers started embedding a higher number of sensors with better calibration and more accurate placement. For immediate response to thermal emergencies in short timescales on-chip sensors are now indispensible to hardware-based mechanisms. This has led to a set of studies, which aim at combining on-chip sensors and application metrics to design and train predictive models, which can be used in a proactive mode. We will review these studies as they relate to our focus more closely. We must highlight that there is a distinct difference between the methods described under post-processing of on-chip sensor data and those that train predictive models with sensor input. The latter often combine application-level and architectural features as indicators of power dissipation and thermal events along with temperature measurements. They may or may not also include power measurements into their model. Higher level features in these methods include hardware performance counters, kernel counters, instruction count, clock frequency, etc.

An early example for training predictive models with temperature sensor input is a regression-based thermal model that has been inserted into the OS's kernel [82]. This model uses input from performance counters at run-time to estimate the thermal state of the system. However, it is trained with online measurements supplied by the on-chip temperature sensors. Another regression-based method utilized only temperature sensor data in an autoregressive moving average (ARMA) model [83]. Sensor readings observed in the past are used to predict a number of steps into the future. The ARMA model is derived from a representative temperature trace of a given workload. At run-time necessary functionality has been provided to collect a new trace and update the model if a significant deviation in the workload characteristics is observed. The recursive least square method (RLSM) has also been used to perform predictions leveraging recent temperature sensor readings as inputs to estimate temperature of a future time step [84].

Performance counters have been often utilized in place of direct power measurements as a good quality proxy, since they closely correlate with processor activity. In turn, processor activity is linearly related to power consumption to a great extent. For instance, a state-space model, similar to the one we discussed in the context of Kalman filtering, has been proposed, where performance counter information and temperature sensor feedback is used to learn the correlation between workload phases and thermal events. This knowledge is used to construct the input matrix of the state-space model instead of power [85]. This model is then used to predict the future temperature for different candidate frequency levels and the performance optimal level that is likely to maintain the peak temperature under a given limit is determined.

Machine learning methods have been a natural medium to investigate temperature prediction methods. A support vector machine (SVM)-based model was trained with thermally relevant features of a system including the measured temperature of a core and its neighbors and the average thermal profiles of tasks assigned to each core for a fixed time interval [86]. Each such feature vector was then labeled as either Hot or NotHot, depending on whether or not the behavior dictated by the corresponding vector resulted in exceeding the safe thermal limit defined for the system. These labels along with the features are then used to train the SVM. Another prominent family of machine learning methods is Reinforcement Learning, which operates on the main principle of a decision mechanism refining itself towards higher reward choices based on what it learns from the response of its environment to the current decisions made. In the context of thermal monitoring, the environment's response is represented with the temperature sensor readings upon a system management decision, for example, task assignment to cores in a many-core chip [87]. Reinforcement learning trained by temperature sensor readings and clock frequency settings has also been used for dynamic voltage scaling and thread assignment [88].

A recent study probed deeper into the suitability of a large number of learning methods to create an online temperature predictor, which uses temperature sensor readings as part of the training phase [89]. For a given system, a large set of features are collected at a time t from variety of sources, including hardware performance counters, temperature sensors, and kernel counters. Some of these features are highly correlated with application characteristics. These features do not change significantly when running the application on different processors and nodes of the system. These time-dependent features are denoted as a vector $\boldsymbol{A}(t)$. The remaining features are more strongly correlated with a node's physical condition. Even while running the same application, these features can vary drastically across nodes depending on each node's cooling conditions, location, and so forth. For example, temperature sensor readings are in this category. This features are called the physical features denoted as a vector $\boldsymbol{P}(t)$. The ultimate goal is to obtain a model for temperature at the next time step i as depicted in (3.9):

$$T(i) = f(\mathrm{A}(i), A(i-1), P(i-1)) \tag{3.9}$$

Learning methods including neural networks, Bayesian networks, linear regression models, multilayer perceptron, Rep Tree, and the Gaussian process have been evaluated. The Gaussian process has been identified as the most successful learning method.

A Gaussian process is a stochastic process of a set of random variables ($X1$, $X2$, ...), where any subset of these variables constitute a joint Gaussian distribution. It has been assumed that any subset of elements of the physical attribute vectors ($\boldsymbol{P}(i_1)$, $\boldsymbol{P}(i_2)$, ..., $\boldsymbol{P}(i_n)$, $\boldsymbol{P}(i_{n+1})$) similarly has a joint Gaussian distribution:

$$(P(i_1), \boldsymbol{P}(i_2), \ldots, \boldsymbol{P}(i_n), \boldsymbol{P}(i_{n+1})) \sim N(0, \mathbf{K}) \tag{3.10}$$

where $K \in \mathbb{R}^{(n+1)\times(n+1)}$ is the covariance matrix. The selection of 0 as the mean of the Gaussian distribution is a common choice. The elements of this matrix are computed by a kernel function k:

$$\begin{aligned} K^{(j,k)} &= k\left(\boldsymbol{X}\left(i_j\right), \boldsymbol{X}\left(i_k\right)\right) \\ \boldsymbol{X}\left(i_j\right) &= \left(\boldsymbol{A}\left(i_j\right), \boldsymbol{A}\left(i_j - 1\right), \boldsymbol{P}\left(i_j - 1\right)\right), \end{aligned} \tag{3.11}$$

The kernel function essentially evaluates the correlation between two samples $X(i_j)$ and $X(i_k)$ and populates the covariance matrix. For a set of observations ($\boldsymbol{P}(i_1)$, $\boldsymbol{P}(i_2)$, ..., $\boldsymbol{P}(i_n)$) and ($\boldsymbol{X}(i_1)$, $\boldsymbol{X}(i_2)$, ..., $\boldsymbol{X}(i_n)$), the model then generates the expected value of $\boldsymbol{P}(i_{n+1})$ (i.e., the prediction) as follows:

$$\mathbb{E}(\boldsymbol{P}(i_{n+1})|X, P, X_{n+1}) = K\left(X_{i_{n+1}}, X\right)\boldsymbol{K}(X, X)^{-1}\boldsymbol{P}, \tag{3.12}$$

where, $P = (P(i_1), \boldsymbol{P}(i_2), \ldots, \boldsymbol{P}(i_n))$,

$$X = (X(i_1), \boldsymbol{X}(i_2), \ldots, \boldsymbol{X}(i_n)).$$

Hence, the temperature prediction model f can be expressed as follows:

$$f = \mathbb{E}(\boldsymbol{P}(i_{n+1})|X, P, (\boldsymbol{A}(i_{n+1}), \boldsymbol{A}(i_{n+1} - 1), \boldsymbol{P}(i_{n+1}))) \tag{3.13}$$

3.4 Guiding thermal management for memory systems with feedback from on-chip temperature sensors

As a consequence of thermal effects becoming prominent system-wide, temperature sensors are now also integrated into components beyond the processor chips. DRAM chips have been equipped with temperature sensors for a few generations. There are also hard disk drives and solid state drives utilizing temperature sensors.

3.4.1 Using sensor feedback for DRAM refresh and write timing optimization

Knowledge of operating temperature for DRAMs has a special relevance. Due to the superlinear dependence of leakage on temperature DRAM cells face the risk of loosing their content if the charge stored in a cell and the corresponding output voltage drops below a limit V_{BL}^{min} that the sense amplifiers of the DRAM can no longer detect. The amount of time it takes for the leakage current to bring the cell voltage V_{cell} below V_{BL}^{min} is defined as the retention time of a DRAM cell. It can be expressed as the summation of time intervals required for each ΔQ_{cell} reduction to occur per ΔV_{cell} step down of the cell voltage:

$$\Delta Q_{cell} = C_{cell} \times \Delta V_{cell}. \tag{3.14}$$

The charge stored in the cell will transition over the range of cell voltage values between V_{DD} and V_{BL}^{min}, which define the correct range of operation. The total retention time can then be expressed as:

$$T_{ret} = \int_{V_{BL}^{min}}^{V_{DD}} \frac{C_{cell} dV_{cell}}{I_{leak}}, \tag{3.15}$$

where C_{cell} is the effective DRAM cell capacitance. This relationship can be approximated as follows:

$$T_{ret} = \frac{C_{cell}\left(V_{DD} - \Delta V_{cell}^{max}\right)}{I_{leak}}. \tag{3.16}$$

The worst-case drop in the DRAM cell voltage ΔV_{cell}^{max}, is a function of the cell capacitance C_{cell}, bitline capacitance C_{BL}, and V_{BL}^{min}, the minimum voltage level that can be detected by the sense amplifier as depicted in (3.17) [90].

$$\Delta V_{cell}^{max} = \frac{V_{DD}}{2} - \frac{C_{cell} + C_{BL}}{C_{cell}} V_{BL}^{min}. \tag{3.17}$$

Plugging ΔV_{cell}^{max} into (3.16) yields:

$$T_{ret} = \frac{C_{cell}\left(\frac{V_{DD}}{2} - V_{BL}^{min}\right) - C_{BL} V_{BL}^{min}}{I_{leak}}. \tag{3.18}$$

Table 3.3 DRAM refresh is a strong function of operating temperature. The amount of current required for refresh also needs to increase with temperature, as does the leakage. As the DRAM capacity increases, the amount of power spent for refresh also increases

	Refresh current @45°C	**Refresh current @85°C**
Low power mobile SDRAM		
64 Mb	100 μA	200 μA
128 Mb	150 μA	250 μA
256 Mb	180 μA	400 μA
DDR3L-SDRAM		
4 GB	180 mA	225 mA

Table 3.3 illustrates the difference between the self-refresh current drawn by an older generation DRAM chip for the mobile market and a newer DDR3 memory chip at different operating temperatures [91, 92].

The refresh rate of the DRAM is calculated such that the retention time described above is not exceeded between consecutive refresh operations of a given cell. We must note that retention time is a strong function of temperature, however, there are other factors that create a variation among different cells in a DRAM even at the ambient temperature. Process variations, data layout, and coupling induced by the access patterns into neighboring cells can cause a wide spread of retention behavior among different cells in a large capacity DRAM [93]. On the other hand, the DRAM refresh rate cannot be set to an arbitrarily high frequency to provide guarantees for the worst case, since this would create a significant overhead in terms of refresh energy. Also, refresh rate impacts the perceived access latency of the DRAM since in most practical implementations the memory device stops servicing data requests from the cores during the refresh period. DRAM manufacturers determine retention time such that the worst-case behavior of a sufficiently large fraction of cells are met to achieve a certain yield within a specified operating temperature range.

DRAM manufacturers have built features into their products to mitigate this problem for many generations of devices. Early products utilized a dedicated temperature sensor chip placed on the DIMM and its temperature reading would be accessed by the system through the serial presence detect (SPD) procedure. Memory modules equipped with the SPD feature essentially contain an EEPROM, where they store various information about their specifications and operating conditions. This information is sent serially from a single EEPROM pin, hence, the "serial" denominator. The data is then transmitted through I^2C bus or its variant SMBus interfaces.

Temperature sensor ICs can be placed on the DRAM chip [94] or on the DIMM board near DRAM chips to track DRAM temperature indirectly. Detection of elevated temperatures can be used for a variety of thermal management actions to mitigate data integrity and reliability concerns for the memory modules. Many

DRAM chips use the sensor readings to define two regions of operations: (i) normal operating temperature range (~0–85°C) and (ii) extended temperature range (~85–95°C) [95]. Older generations of DRAM devices would only be rated up to 95°C for commercial applications. Military and space applications and other special domains could utilize memory chips rated for higher temperatures.

One of the most common methods is called the auto temperature compensated self refresh or adaptive refresh [96–98]. Based on the feedback received from the built-in temperature sensors the refresh rate of the memory chip is adjusted to adapt to increasing thermal-induced leakage. The double data rate (DDR) standard established a maximum of 64 ms separation between two consecutive refresh operations to the same location commercial DRAM chips. This is usually divided into several smaller refresh operations to subdivisions. For instance, within 64 ms a memory ranks of 8192 partitions can be refreshed if each partition is refreshed every 7.8 μs. While this refresh period is imposed for the normal operating range, the standard requires to halve the period to 32 ms (3.9 μs per partition) for temperatures above 85°C. In emerging 3D stacked DRAM technologies the impact of thermal effects becomes even more significant. A recent memory module designed as a 3D stacked chip introduced a third refresh period of 8 ms for the temperature range of 95–115°C [99].

Another timing parameter which has a dependence on temperature is the write recovery time, defined as the amount of time required to write data into a DRAM cell. The write recovery time has in fact an inversed relationship with temperature. The amount of time required for reliable operation increases with decreasing operating temperature [100]. Use of an on-chip temperature sensor to monitor the DRAM temperature has been proposed in response to this observation. The length of the write recovery time period can be modulated with the specific timing of the precharge signal following the last write. It has been proposed to adjust the write recovery period, by extending it if the DRAM temperature drops below a pre-characterized threshold. Since this timing parameter has a direct impact on the overall DRAM latency, it cannot be set to an arbitrarily large value for all cases. It is therefore crucial to leverage dynamic monitoring support from on-chip sensors to reduce the period back to a lower value when the temperature threshold has been exceeded.

3.4.2 Using sensor feedback in architectural optimizations for DRAMs

The two usage cases for dynamic thermal monitoring described above are related directly to the performance and power consumption of the DRAM. In addition, high operating temperatures would be undesirable for reliability concerns.

In order to mitigate the high operating temperatures, the activity on the DRAM can be throttled similar to processor chips. A common method for this is to actually throttle the bandwidth of the bus connecting the cores and the DRAM or adaptively gating cores to reduce their access intensity [101, 102]. By adjusting the rate of arrival for memory requests, the activity of the DRAM can be modulated. In addition, the temperature sensor feedback from the memory subsystem can be used to control active cooling, such as fan speed, similar to what is practiced with processor chips.

3.4.3 Using sensor feedback for thermal management in hard drives

Thermal effects are important for hard disks from the reliability point of view. High operating temperatures can cause off-track errors due to expansion in the disk stack and actuators under high thermal stress, which in turn can impact the performance of the disk and even cause fatal errors. Disk drives in consumer products have started utilizing embedded temperature sensors in their latest generations. The system software can access these sensor measurements through device interfaces. There are products that incorporate feedback from a temperature sensor to a cooling fan that reacts to hard drive temperature change in real-time.

A detailed analysis on the roadmap for disk drives from a thermal perspective established that the rate of increase in internal data rates and capacities cannot be sustained under the assumption that the disk drives' operating temperatures would not elevate proportionally. The projections on the performance and density parameters of this study indicate that the resulting thermal state of hard disks will be extremely close to if not exceeding the worst-case temperature ratings [103]. However, the manufacturers can only meet their design targets under the assumption of average case temperatures. This strongly indicates that DTM, much like other system components reviewed in this chapter so far, would be an indispensable tool to continue to meet performance targets under a safe thermal envelope for hard disk drives. A possible method for performance throttling in response to a thermal trigger is dynamically regulating the amount of requests sent to the disk for a cooling down period, before ramping up the activity again upon a thermal sensor's signal.

A follow up study investigated specific knobs for thermal management of disk drives [104]. Through construction of microbenchmarks applications have been characterized in terms of their thermal impact. Time separation between seeks operations in a workload has been found to impact the opportunities to utilize performance throttling for DTM of hard disks. In another study, a multi-speed disk drive has been proposed for mitigating thermal emergencies [104]. By modulating the rotational speed of the platter dynamically, a thermal management effect akin to frequency throttling in processors is achieved.

3.5 A closer look at the control systems for thermal management

Embedded temperature sensors play a vital role in DTM in almost all of today's high-performance system components, ranging from processor cores, to disk drives and from network-on-chips to co-processors and accelerators. At the other end of the control loop, advancements in circuit design techniques enable highly responsive and increasingly finer grained control knobs to modulate voltage, frequency, communication bandwidth, etc. Especially in sophisticated run-time control methods, these two agents are connected by another critical component, which is the control unit. As we conclude this chapter on the role and various uses of thermal sensor feedback in DTM, it will provide useful insight to review a select set of prominent control systems widely used in dynamic management. In future systems, co-design and co-optimization of these three major agents might draw further attention. In the following, four major types of control mechanism that were encountered in the treatment of DTM will be described with a brief background on the theoretical aspects of each mechanism.

3.5.1 Closed loop (feedback) controller

A closed loop controller uses feedback from the state of the system under control. In thermal management this corresponds to temperature sensor measurements. The control actions are usually executed at a fixed frequency. For instance, temperature sensors are sampled at a fixed rate by the thermal monitoring circuits and fed to the control mechanism, either as a complete set of values or after some pre-processing to determine the highest value. There are four major components of the closed loop control system: (i) the controller *C*; (ii) sensor *S*; (iii) device under control *D;* and (iv) reference setting *R*. Figure 3.11 illustrates these basic components operating in a feedback loop.

The reference setting for the controller could be the temperature threshold to trigger performance throttling. The error evaluation block of the controller computes the difference between the value reported by the sensor feedback and the reference setting. The controller mechanisms usually operate in an attempt to minimize this disagreement term $e = R - S$ by generating a proper response via its actuators to drive the device under control with a new set of inputs u. The device under control will respond to the new input and generate the output y, which will be monitored by the sensors and placed into the feedback loop again.

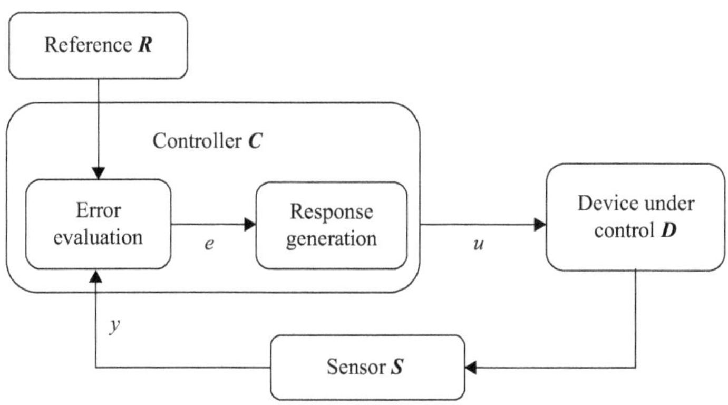

Figure 3.11 A typical feedback loop in a closed loop control system

3.5.1.1 Proportional (integral) (derivative) controllers

Transfer functions of control mechanisms can be categorized according to the processing they apply on the error term e. The simplest case would be a Proportional Controller, which generates a response that is proportional to e by a constant factor. It is often challenging to determine the best choice for the constant coefficient to achieve a control mechanism without overshoot/undershoot or oscillation. This basic scheme is enhanced by involving the integration of the error term over time. This helps the convergence by responding to the errors of small magnitude can persist in the system for a long duration time via integration. The resulting output function of the controller can be expressed as follows:

$$u(t) = K_P e(t) + K_I \int e(t)dt \quad \text{where } e(t) = y(t) - R \tag{3.19}$$

The output function of a controller is often expressed in frequency domain in the form of a Laplace transform as shown in (3.20), where the time-dependent functions $e(t)$, $u(t)$, and $y(t)$ governing the control system are transformed into function of a complex frequency variable s.

$$U(s) = K_P + \frac{K_I}{s} \tag{3.20}$$

The resulting controller is the proportional integral (PI) controller. Another extension to this model is to also incorporate the derivative term of the error term to provide smoothing of the response and react to changes which may indicate that the controller would otherwise overshoot or undershoot. The resulting system is called a proportional-integral-derivative (PID) controller as shown in (3.21).

$$u(t) = K_P e(t) + K_I \int e(t)dt + K_D \frac{de(t)}{dt} \quad \text{and} \quad U(s) = K_P + \frac{K_I}{s} + K_D s \tag{3.21}$$

For DTM closed loop controllers has been proposed to control the clock's duty cycle [105] and selection of (voltage, frequency) setting [62, 63]. If DTS outputs are fed into the control system, then the integration also needs to be approximated in discrete form. In that case, integration can be computed as a summation of error terms computed for each sampling interval multiplied by the length of the interval. A history of past error terms would then need to be stored by the controller. This can be performed by the dedicated microcontrollers often used for thermal monitoring and management. If the controller is receiving the analog temperature sensor output as a voltage or current signal, then, an operational amplifier circuit can be utilized to compute the integration component. Another important issue is determining the values of the constant coefficients. By setting a subset of these three major coefficients, the controller can be configured to be only proportional to error, proportional derivative of error signal, or proportional to the integration or error signal (summation of error signals over discrete samples). In one of the realizations all three control action components were utilized [105], while the other omitted the derivative component due to its insignificant contribution based on empirical observations [62] or due to its susceptibility to noise [63].

While PID controllers are effective in many settings, they possess non-idealities. Naturally, the fact that the gain parameters are fixed introduces some limitations and the tuning of these gain parameters will impact the ultimate success of the controller. The device under control as well as the sensors may also introduce gain factors into their response. If these are not characterized well or minimized to be negligible the effectiveness of the controller could be limited.

Another aspect calling for special attention of PID controllers is the integral windup phenomenon. Integration component of the error term can accumulate into a large value over time even though the individual error values of each time step is small. The controller would drive the actuators into their upper limit (e.g., higher (voltage, frequency) setting)) in such a case causing a saturation of the actuator. If such a long epoch of feedback samples with positive error terms is followed by another where the thermal sensors indicate an overshoot in thermal state by a change in the sign of the error term, the controller cannot adapt soon enough. It would require a period of time to "wind down" to reduce the integral term. This may cause the thermal state to exacerbate in the meanwhile, since the controller will not engage throttling. The opposite can also occur, where during a prolonged period of high activity and peak temperatures the controller will react by engaging the actuator to quickly drop down to maximal throttling level. However, due to the physical properties of the device under control, the thermal response to minimizing the performance level will not occur instantaneously. It is limited by the inherent time constant of the hardware. During this period, which is likely to be longer than the sampling period of the temperature sensors, the integration component would again accumulate a large quantity of this hidden error. When the device finally cools down, the accumulated integration component will not diminish instantaneously causing the system to operate at a lower performance point than necessary. Integral windup can be mitigated by a modification on the controller. One option is to monitor the saturation events at the actuator and disable the integration component after detecting saturation. Also, in the discrete implementation of the integration using digital sensor outputs, the integration can be limited in terms of the number of discrete steps involved, which will in turn limit the degree of windup during integration.

Finally, the derivative component is sensitive to noise in the feedback measurement. It can be mitigated by pre-filtering the sensor feedback with a low-pass filter to remove high frequency components in the noise.

There are also factors beyond the closed loop control theory that influence the resulting controller's ability in achieving a dynamic thermal control mechanism which maintains the system under a thermal envelope while maximizing its power efficiency. The actuator latency, that is, the delay in implementing the new (voltage, frequency) setting indicated by the controller is a limitation beyond the scope of the controller itself. Similarly, the physical nature of the materials involved will only allow to cool down in response to throttling with a certain time constant. Unless material properties are changed or other external cooling is supplemented, the closed loop controller in charge of determining the setting of the hardware-based performance knobs can influence the thermal output within a limit. That is the reason why the hardware-based DTM mechanisms also implement an emergency shutdown mode. Having said that, it comes at no surprise that feedback control mechanisms have also been utilized in various outer loops of thermal management to trigger cooling systems, such as cooling fan speeds.

3.5.1.2 Linear quadratic regulator algorithm for configuring feedback controllers

The state-space representation was discussed earlier in this chapter as a means to translate imprecise or incomplete temperature sensor measurements into a finer thermal map for refined feedback. The transfer function of the DTM controller can be expressed similarly in this manner and a linear feedback control system can be described through this representation. The state-space representation of a linear system in time discrete form can be expressed with (3.22) and (3.23):

$$x[n+1] = Ax[n] + Bu[n] + N\phi[n] \tag{3.22}$$

$$y[n] = Cx[n] + Du[n] + \sigma[n] \tag{3.23}$$

where, $x[n]$ represents the state of the device under control (i.e., temperature), $u[n]$ represents the input to the system (i.e., frequency, voltage), and $y[n]$ represents the system's output. Each equation also contains an error term to represent inaccuracies in modeling the system (3.22) and noise in the feedback sent to the controller (3.23). This linear feedback system can take the form of a proportional gain system for thermal management. For this controller, there exists a proportional gain factor K_p, which minimizes a performance metric (cost function) of the feedback system. The objective of the feedback system is defined with a cost function $J(u)$ representing the reward/overhead of an action taken by the controller, which depends on the system's state $x[n]$ and controller's output $u[n]$.

An algorithmic approach to establishing the cost function and identifying the gain factor that minimizes it is called the linear quadratic regulator (LQR) algorithm. An application of LQR to configure a DTM control system has been proposed [106]. In this particular system, the operating frequency of a processor is regulated to drive it at an optimal performance-power tradeoff point, while at the same time maintaining the thermal state under a threshold.

The objective function $J(u)$ corresponding to a general state-space representation (after a simplification by removing the noise terms) is as follows

$$J(u) = \sum_{n=1}^{\infty} \left(x[n]^T Qx[n] + u[n]^T Ru[n] \right) \tag{3.24}$$

For instance, the thermal network representation of the device under control can be used in place of the Q matrix so that first term denotes the thermal consequence of the controller‘s action; and the matrix R in the second term, can represent the power consumption resulting from a frequency setting supplied through the controller's input signal $u[n]$. The formulation is referred to as the infinite horizon discrete time LQR, when the performance metric is evaluated through the entire timescale. If it is constrained to a finite set of time steps, the formulation is called the finite horizon discrete time LQR.

For the case of thermal management, the optimization problem for this cost function is subject to constraints that signify minimum and maximum bounds for allowed clock frequencies, an equality that associates clock frequency with an actual value for $u[n]$, and the two fundamental equations describing the state-space representation of the system (see (3.22) and (3.23)). The optimal gain factor K_p used by the controller's feedback loop minimizes this cost function and it can be derived as:

$$K_P = \left(B_{eq}^T S B_{eq} + R\right)^{-1} \left(B_{eq}^T S A_{eq}\right) \tag{3.25}$$

The matrices B_{eq} and A_{eq} are equivalent matrices to B and A in the state-space representation, adjusted for the specific sampling period of the thermal management system. S is a positive definite matrix computed by solving the discrete algebraic Riccati equation.

Ultimately, the linear feedback control system would execute a policy of:

$$u[n] = Reference\_FREQ - K_p x[k] \tag{3.26}$$

3.5.2 Stochastic control

The uncertainty in the observations obtained from temperature sensor measurements and the additional noise in conversion circuits and communication interfaces renders the input to the controller to be of stochastic nature. In recognition of this, stochastic control mechanisms have been applied to the problem of DTM. Stochastic control essentially views the observed state as a random variable with some assumed probability distribution. Furthermore, in the case of thermal management, the variables are discrete time variables. The objective of the controller then turns into minimization of the expected value of the error between the desired state and the observed state, for example, temperature limit and observed temperature.

One implementation of stochastic control is through the Markovian process, which can be viewed as an extension to a Markov chain. A Markov chain defines a stochastic process comprised of a finite set of n states ($s_1, \ldots, s_n$). A system starting at a given state s_i, will have a predetermined probability to making a transition to another state in this set according to a known probability distribution. A Markovian process extends this system by introducing two new concepts: actions (or choices) and rewards (or cost).

In each state, a controller can now have multiple choices and in response to a specific choice taken, the process will transition into a new state with a certain probability. In this case, the transition probability (TP) is not only determined by a fixed distribution globally, but also by the specific action taken. Still, the transition is independent of past states and hence, also past choices. Furthermore, each choice is associated with a reward.

TP function describing the system will dictate the probability of transitioning from a state s_i at any time step t to state s_j under action a as follows:

$$TP(s_j, a, s_i) = P(s(t+1) = s_j \,|a(t) = a \ \vee s(t) = s_i), \tag{3.27}$$

The task of the controller is to implement a policy that will move the process towards a next state with an action that minimizes the associated expected cost (or maximize reward). One approach to solving this problem is to iterate over all possible combinations of states, actions, and observations to calculate their respective cost and move towards the smallest cost and continue computing policy choices and their resulting cost over a number of iterations until no improvement is observed. The cost is always computed as a cumulative quantity across possible states and actions with their respective probability weight and also finally adjusted with a so-called discount factor γ (a value that is usually set close to 1).

In its application to DTM, states would correspond to the thermal state of the device under control and actions correspond to discrete settings of the (voltage, frequency) values. Such a process model will need to involve one simplifying assumption to maintain computational efficiency. Thermal states cannot be defined at an arbitrarily fine level of granularity for the entire operating range, since this would lead to an excessive number of states. One realization proposed to divide the operating temperature range into three sub-ranges [68] (e.g., a low-temperature range: ambient air temp. $\leq T_{\text{sensor}} < T_{cool}$, a medium temperature range: $T_{cool} \leq T_{sensor} < T_{medium}$, and a high-temperature range defining the threshold for triggering aggressive throttling: $T_{medium} \leq T_{sensor} \leq T_{limit}$). The overall flow of the control process is illustrated in Figure 3.12.

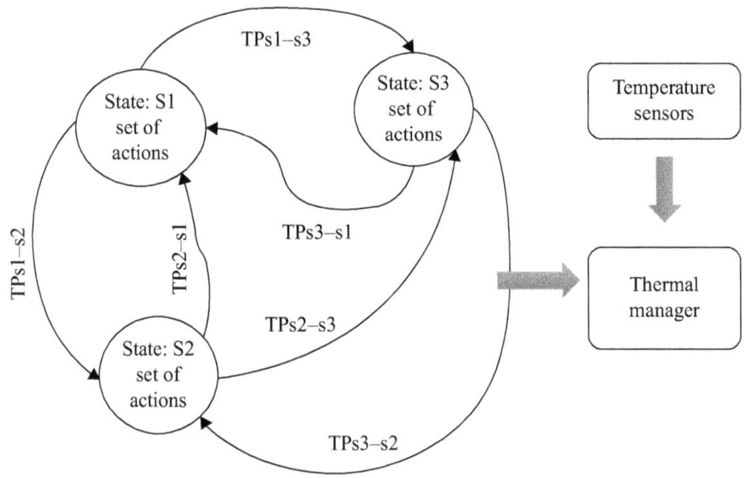

Figure 3.12 The Markovian process model introduces TP between states. The thermal manager considers possible available actions at each state and executes a policy to enable a transition. The consequence (reward or cost) of an action is evaluated using the temperature sensor feedback

There can be variants to the basic formulations of the stochastic controller. One variant that is relevant to thermal management stems from the issue of observability. This can occur, when the thermal management controller is not able to accurately observe the direct consequence of a selected action on the thermal state due to interference of external dynamic factors. For instance, before the thermal response to a recent switch into a new frequency and voltage setting takes place (and gets to be observed by the temperature sensors), a change in the workload may lead the OS to intervene and place the device into an idle mode, drastically shifting the thermal response. In order to have the system conform to the Markovian process assumption, observability can be re-introduced by translating each state into a belief state, which is updated dynamically upon each state transition. The belief state essentially allows a definition of state as a probabilistic combination of all possible states with the total probability equal to 1.

Reinforcement learning is another avenue of stochastic control and optimization that is drawing an increasing amount of attraction in recent years for a variety of power and performance optimization methods targeting high-performance processors and systems on chip [87, 88, 107]. This method also assumes that a Markov process is governing the underlying system to be controlled. However, it does not require a priori knowledge of the distribution probabilities of the process. Instead, consequences of actions are monitored and the process "learns" over time which actions led to good (desirable) results and which did not.

3.5.3 Model predictive control

Model predictive control (MPC) is another attractive control concept, which involves prediction of the outcome for a decision that will be made at the current time step. In addition to sensor feedback, this method also requires a model of the system and an appropriate light-weight solver (simulator) for that model to perform online predictions on its actions. The aforementioned solver in fact also includes the entire optimization problem that the controller is in charge of pursuing. For instance, in the context of thermal management, the dynamic controller aims to operate the device under control as close to the maximum allowed power consumption as possible (indicating highest performance at highest frequency) while the resulting thermal state should not exceed a threshold. In this case, the system's thermal model would be combined with the power-performance indicative objective function and this resulting optimization problem would be solved by the solver embedded in the MPD engine. The resulting solution will indicate the maximum allocation of the input resources into the system (e.g., power budget in a chip) that will not exceed the safe temperature limit, if the predicted temperature agrees with the eventual temperature resulting from this decision. One realization of the MPC concept for DTM has defined the overall control system in three major blocks: (i) a frequency-to-power block, which determines the maximum allowable power budget under the current operating frequency and workload; (ii) controller block, which contains the observer (i.e., temperature sensors) and the optimization solver; and (iii) a power-to-frequency block which receives the specific amount of power determined by the controller to be consumed in the next execution interval and converts it to the actual clock frequency setting to drive the clock circuitry [108].

Another MPC-based method proposes an explicit computation approach for the optimization solver to mitigate computational complexity [109]. Their main observation is that the state space can be divided into well-defined regions, where for each region an optimization problem can be computed offline and stored in a look-up table.

References

[1] Sanchez, H., et al., Thermal management system for high performance PowerPC(TM) microprocessors. IEEE Compcon 97, Proceedings, 1997: pp. 325–330.

[2] Floyd, M.S., et al., System power management support in the IBM POWER6 microprocessor. IBM Journal of Research & Development, 2007. **51**(6): pp. 733–746.

[3] Floyd, M., et al., Introducing the adaptive energy management features of the power7 chip. IEEE Micro, 2011. **31**(2): pp. 60–74.

[4] Floyd, M., et al., Adaptive energy-management features of the IBM POWER7 chip. IBM Journal of Research and Development, 2011. **55**(3).

[5] Pant, M., *Microprocessor Power Impacts*, in *Great Lakes VLSI Symposium*. 2010.

[6] Dorsey, J., et al., *An integrated quad-core opteron processor*. in *Solid-State Circuits Conference, 2007. ISSCC 2007. Digest of Technical Papers. IEEE International*. 2007.

[7] Bendali, A. and Y. Audet, A 1-V CMOS current reference with temperature and process compensation. IEEE Transactions on Circuits and Systems I-Regular Papers, 2007. **54**(7): pp. 1424–1429.

[8] Ituero, P., M. Lopez-Vallejo, and C. Lopez-Barrio, A 0.0016 mm(2) 0.64 nJ leakage-based CMOS temperature sensor. Sensors, 2013. **13**(9): pp. 12648–12662.

[9] Remarsu, S. and S. Kundu, *On Process Variation Tolerant Low Cost Thermal Sensor Design in 32nm CMOS Technology*. Glsvlsi 2009: Proceedings of the 2009 Great Lakes Symposium on VLSI, 2009: pp. 487–492.

[10] Remarsu, S., *On Process Variation Tolerant Low Cost Thermal Sensor Design*, in *Electrical and Computer Engineering*. 2014. University of Massachusetts Amherst: Amherst, MA. p. 72.

[11] Duarte, D.E., et al., *Advanced Thermal Sensing Circuit and Test Techniques Used in a High Performance 65nm Processor*. in *Islped'07: Proceedings of the 2007 International Symposium on Low Power Electronics and Design*, 2007: pp. 304–309.

[12] Pertijs, M.A.P. and J.H. Huijsing, *Transistor temperature measurement for calibration of integrated temperature sensors*. in *Instrumentation and*

Measurement Technology Conference, 2002. IMTC/2002. Proceedings of the 19th IEEE. 2002. IEEE.

[13] Long, J., et al., Thermal monitoring mechanisms for chip multiprocessors. ACM Transactions on Architecture and Code Optimization, 2008. **5**(2): pp. 9:1–9:33.

[14] Chen, S., et al., *Floorplanning with consideration of white space resource distribution for repeater planning*. in *Quality of Electronic Design, 2005. ISQED 2005. 6th International Symposium on*. 2005. IEEE.

[15] Gröpl, C., et al., *Approximation algorithms for the Steiner tree problem in graphs*, in *Steiner trees in industry*. 2001. Springer. pp. 235–279.

[16] Lin, G.H. and G.L. Xue, On the terminal Steiner tree problem. *Information Processing Letters*, 2002. **84**(2): pp. 103–107.

[17] Long, J. and S.O. Memik, *Optimization of the bias current network for accurate on-chip thermal monitoring*. in *Design, Automation & Test in Europe Conference & Exhibition (DATE), 2010*. 2010. IEEE.

[18] Memik, S.O., et al., Optimizing thermal sensor allocation for micro-processors. *IEEE Transactions on Computer-Aided Design of Integrated Circuits and Systems*, 2008. **27**(3): pp. 516–527.

[19] Ituero Herrero, P., *On-Chip Thermal Monitoring: Design, Placement and Interconnection of Temperature Sensors*. 2012. Telecomunicacion, Universidad Politecnica de Madrid, Madrid, Spain.

[20] Szekely, V., M. Rencz, and B. Courtois, Tracing the thermal behavior of ICs. IEEE Design & Test of Computers, 1998. **15**(2): pp. 14–21.

[21] IEEE, *IEEE Std 1149.1-2001, IEEE Standard Test Access Port and Boundary-Scan Architecture*. 2001.

[22] Ingelsson, F.G.Z.U., G. Carlsson, and E. Larsson, *Reusing and Retargeting On-Chip Instrument Access Procedures in IEEEP1687*. 2012.

[23] *Serial Vector Format Specification* http://www.asset-intertech.com/support/svf.pdf 1999.

[24] *TruTherm Technology Presentation*, in *Literature Number: SNIA013*. 2006, Texas Instruments.

[25] *Remote/Local Temperature Sensor with SMBus Serial Interface*, in http://www.maximintegrated.com/en/products/analog/sensors-and-sensor-interface/MAX1617.html, M. Integrated, Editor. 2015.

[26] Lopez-Buedo, S. and E. Boemo, *Making visible the thermal behaviour of embedded microprocessors on FPGAs: A progress report*. in *Proceedings of the 2004 ACM/SIGDA 12th International Symposium on Field Programmable Gate Arrays*. 2004. ACM.

[27] Mondal, S., R. Mukherjee, and S.O. Memik, *Fine-grain thermal profiling and sensor insertion for FPGAs*. 2006 IEEE International Symposium on Circuits and Systems, **Vols 1–11**, Proceedings, 2006: pp. 4387–4390.

[28] Sundararajan, P., et al., *Thermal characterization and optimization in platform FPGAs*. in *Proceedings of the 2006 IEEE/ACM International Conference on Computer-aided Design*. 2006. ACM.

[29] Velusamy, S., et al., *Monitoring temperature in FPGA based SoCs*. in *Computer Design: VLSI in Computers and Processors, 2005. ICCD 2005. Proceedings. 2005 IEEE International Conference on*. 2005. IEEE.

[30] *On-Chip Peripheral Bus (OPB)*: http://www.xilinx.com/support/documentation/ip_documentation/opb_v20.pdf

[31] Madduri, S., *MNoC: A Network on Chip for Monitors*. 2008. University of Massachusetts Amherst, Amherst, MA, USA.

[32] Zhao, J., et al., A dedicated monitoring infrastructure for multicore processors. Very Large Scale Integration (VLSI) Systems, IEEE Transactions on, 2011. **19**(6): pp. 1011–1022.

[33] Phanibhushana, B., et al., *Towards efficient on-chip sensor interconnect architecture for multi-core processors*. in *SoC Design Conference (ISOCC), 2010 International*. 2010. IEEE.

[34] Vaddina, K.R., et al., *On-line distributed thermal sensing and monitoring of multicore systems*. in *Norchip, 2008*. 2008. IEEE.

[35] *Intel® Pentium® 4 Processor with 512-KB L2 Cache on 0.13 Micron Process and Intel® Pentium® 4 Processor Extreme Edition Supporting Hyper-Threading Technology - Datasheet*. 2004, Intel, Inc.

[36] *Application Note: Using Thermal Diodes in the PowerPC970MP® Processor*. 2006, IBM.

[37] Rotem, E., et al., *Analysis of thermal monitor features of the Intel Pentium M processor*. in *TACS Workshop at ISCA-31*. 2004.

[38] *AMD Athlon™ 64 Processor Power and Thermal Data Sheet*. 2006, Advanced Micro Devices.

[39] *Intel® Atom™ Processor D400 and D500 Series Thermal/Mechanical Specifications and Design Guidelines*. 2009, Intel, Inc.

[40] Michaud, P., et al., *A study of thread migration in temperature-constrained multicores*. ACM Transactions on Architecture and Code Optimization (TACO), 2007. **4**(2): p. 9.

[41] *PowerPC 740TM PowerPC 750TM RISC Microprocessor User's Manual*. 1999, IBM.

[42] Hamilton, D., *Application Note: Migrating from IBM 750GX to MPC7447A*. 2005. Freescale Semiconductor.

[43] Poirier, C., et al., *Power and temperature control on a 90nm Itanium®-family processor*. in *Solid-State Circuits Conference, 2005. Digest of Technical Papers. ISSCC. 2005 IEEE International*. 2005. IEEE.

[44] McGowen, R., et al., Power and temperature control on a 90-nm Itanium family processor. IEEE Journal of Solid-State Circuits, 2006. **41**(1): pp. 229–237.

[45] Fischer, T., et al., *A 90nm variable-frequency clock system for a power-managed Itanium®-family processor*. in *Solid-State Circuits Conference, 2005. Digest of Technical Papers. ISSCC. 2005 IEEE International*. 2005. IEEE.

[46] Wamhoff, J.-T., et al., *The TURBO diaries: Application-controlled frequency scaling explained*. in *Proceedings of the 2014 USENIX conference on USENIX Annual Technical Conference*. 2014. USENIX Association.

[47] Zyuban, V., et al., IBM POWER8 circuit design and energy optimization. IBM Journal of Research and Development, 2015. **59**(1): pp. 9:**1**–**9**:16.

[48] Berktold, M. and T. Tian, *White Paper: CPU Monitoring with DTS/PECI*. 2010. Intel Corporation, Portland, OR, USA.

[49] *Dual-Core Intel® Xeon® Processor 5100 Series Thermal/Mechanical Design Guidelines*. 2006. Intel Corporation.

[50] *Understanding Power Management of Intel® Processors for Mil/Aero Applications*. 2011. Curtiss Wright Controls Embedded Computing.

[51] *Intel® Xeon Phi™ Coprocessor x100 Product Family Datasheet*. 2015. Intel Corporation.

[52] Joonho, K., C. Sung Woo, and S. Kevin, *Recent thermal management techniques for microprocessors*. ACM Computing Surveys, 2012. **44**(3): pp. 1–42.

[53] Brooks, D. and M. Martonosi, *Dynamic thermal management for high-performance microprocessors*. in *High-Performance Computer Architecture, 2001. HPCA. The 7th International Symposium on*. 2001. IEEE.

[54] Thidapat, C., X.S. Hu, and P.D. Robert, *Online work maximization under a peak temperature constraint*. in *Proceedings of the 2009 ACM/IEEE International Symposium on Low Power Electronics and Design*. 2009. ACM, San Fancisco, CA, USA.

[55] Huang, M., et al., *A framework for dynamic energy efficiency and temperature management*. in *Proceedings of the 33rd Annual ACM/IEEE International Symposium on Microarchitecture*. 2000. ACM.

[56] Mukherjee, R. and S.O. Memik, *Physical aware frequency selection for dynamic thermal management in multi-core systems*. IEEE/Acm International Conference on Computer-Aided Design, Digest of Technical Papers, Iccad, 2006: pp. 715–720.

[57] Swaroop, G., et al., *O2C: Occasional two-cycle operations for dynamic thermal management in high performance in-order microprocessors*. in *Proceedings of the 2008 International Symposium on Low Power Electronics & Design*. 2008. ACM, Bangalore, India.

[58] Srinivasan, M., et al., *Temperature control of high-performance multi-core platforms using convex optimization*. in *Proceedings of the Conference on Design, Automation and Test in Europe*. 2008. ACM, Munich, Germany.

[59] Dongkeun, O., et al., *Runtime temperature-based power estimation for optimizing throughput of thermal-constrained multi-core processors*. in *Proceedings of the 2010 Asia and South Pacific Design Automation Conference*. 2010. IEEE Press, Taipei, Taiwan.

[60] Zhijian, L., et al., Improved thermal management with reliability banking. Micro, IEEE, 2005. **25**(6): pp. 40–49.

[61] Bing, S., Z. Yufu, and S. Ankur, *Dynamic thermal management for single and multicore processors under soft thermal constraints*. in *Proceedings of the 16th ACM/IEEE International Symposium on Low Power Electronics and Design*. 2010. ACM, Austin, Texas, USA.

[62] Donald, J. and M. Martonosi, *Techniques for multicore thermal management: Classification and new exploration*. in *ACM SIGARCH Computer Architecture News*. 2006. IEEE Computer Society.

[63] Kadin, M., S. Reda, and A. Uht, *Central vs. distributed dynamic thermal management for multi-core processors: Which one is better?* in *Proceedings of the 19th ACM Great Lakes Symposium on VLSI*. 2009. ACM.
[64] Jonathan, A.W. and H.A. David, Addressing thermal nonuniformity in SMT workloads. ACM Transactions on Architecture and Code Optimization, 2008. **5**(1): pp. 1–28.
[65] Changyun, Z., et al., Three-dimensional chip-multiprocessor run-time thermal management. Computer-Aided Design of Integrated Circuits and Systems, IEEE Transactions on, 2008. **27**(8): pp. 1479–1492.
[66] Chiueh, H., J. Draper, and J. Choma, A dynamic thermal management circuit for system-on-chip designs. Analog Integrated Circuits and Signal Processing, 2003. **36**(1–2): pp. 175–181.
[67] Lee, W., K. Patel, and M. Pedram, *Dynamic thermal management for mpeg-2 decoding*. in *Proceedings of the 2006 International Symposium on Low Power Electronics and Design*. 2006. ACM.
[68] Hwisung, J. and M. Pedram, *Stochastic dynamic thermal management: A Markovian decision-based approach*. in *Computer Design, 2006. ICCD 2006. International Conference on*. 2006.
[69] Kalman, R., A new approach to linear filtering and prediction problems. Transactions of the ASME - Journal of Basic Engineering, 1960. **D**(**82**): pp. 35–45.
[70] Sharifi, S., L. Chunchen, and T.S. Rosing, *Accurate temperature estimation for efficient thermal management.* in *Quality Electronic Design, 2008. ISQED 2008. 9th International Symposium on.* 2008.
[71] Qian, H., et al., Cyber-physical thermal management of 3D multi-core cache-processor system with microfluidic cooling. Journal of Low Power Electronics, 2011. **7**(1): pp. 110–121.
[72] Yufu, Z. and S. Ankur, *Adaptive and autonomous thermal tracking for high performance computing systems*. in *Proceedings of the 47th Design Automation Conference*. 2010. ACM, Anaheim, California.
[73] Yufu, Z. and A. Srivastava. *Leakage-aware Kalman filter for accurate temperature tracking*. in *Green Computing Conference and Workshops (IGCC), 2011 International*. 2011.
[74] Zjajo, A., N. van der Meijs, and R. Van Leuken, Dynamic thermal estimation methodology for high-performance 3-D MPSoC. Very Large Scale Integration (VLSI) Systems, IEEE Transactions on, 2014. **22**(9): pp. 1920–1933.
[75] Hwisung, J. and M. Pedram, *A stochastic local hot spot alerting technique*. in *Design Automation Conference, 2008. ASPDAC 2008. Asia and South Pacific*. 2008.
[76] Odabasioglu, A., M. Celik, and L.T. Pileggi, PRIMA: Passive reduced-order interconnect macromodeling algorithm. 1997 IEEE/Acm International Conference on Computer-Aided Design - Digest of Technical Papers, 1997: pp. 58–65.
[77] Zhang, Y. and A. Srivastava, Accurate temperature estimation using noisy thermal sensors for Gaussian and non-Gaussian cases. Very Large Scale *Integration (VLSI) Systems, IEEE Transactions on*, 2011. **19**(9): pp. 1617–1626.

[78] Zhang, Y., A. Srivastava, and M. Zahran, On-chip sensor-driven efficient thermal profile estimation algorithms. ACM Transactions on Design Automation of Electronic Systems (TODAES), 2010. **15**(3): p. 25.

[79] Zhang, Y. and A. Srivastava, *Statistical characterization of chip power behavior at post-fabrication stage*. in *Green Computing Conference and Workshops (IGCC), 2011 International*. 2011. IEEE.

[80] Zhang, Y., B. Shi, and A. Srivastava, *A statistical framework for designing on-chip thermal sensing infrastructure in nano-scale systems*. in *Proceedings of the 19th International Symposium on Physical Design*. 2010. ACM.

[81] Zjajo, A., N.V.D. Meijs, and R.V. Leuken, Adaptive thermal monitoring of deep-submicron CMOS VLSI circuits. Journal of Low Power Electronics, 2013. **9**(4): pp. 1–11.

[82] Kumart, A., et al., HybDTM: A coordinated hardware-software approach for dynamic thermal management. in *43rd Design Automation Conference, Proceedings* 2006, 2006: pp. 548–553.

[83] Coskun, A.K., T.S. Rosing, and K.C. Gross, Utilizing predictors for efficient thermal management in multiprocessor SoCs. Computer-Aided Design of Integrated Circuits and Systems, IEEE Transactions on, 2009. **28**(10): pp. 1503–1516.

[84] Yeo, I., C.C. Liu, and E.J. Kim, *Predictive dynamic thermal management for multicore systems*. in *Proceedings of the 45th Annual Design Automation Conference*. 2008. ACM.

[85] Cochran, R. and S. Reda, *Consistent runtime thermal prediction and control through workload phase detection*. in *Proceedings of the 47th Design Automation Conference*. 2010. ACM.

[86] Yun, B., K.G. Shin, and S. Wang, *Predicting thermal behavior for temperature management in time-critical multicore systems*. in *Real-Time and Embedded Technology and Applications Symposium (RTAS), 2013 IEEE 19th*. 2013. IEEE.

[87] Lu, S.J., R. Tessier, and W. Burleson, *Reinforcement Learning For Thermal-Aware Many-Core Task Allocation*, in *Great Lakes VLSI Symposium*. 2015.

[88] Das, A., et al., *Reinforcement learning-based inter- and intra-application thermal optimization for lifetime improvement of multicore systems*. in *Design Automation Conference (DAC), 2014 51st ACM/EDAC/IEEE*. 2014.

[89] Zhang, K., et al., *Minimizing Thermal Variation Across System Components*, in *IEEE International Parallel & Distributed Processing Symposium*. 2015.

[90] Thoziyoor, S., et al., *CACTI 5.1*. 2008, Hewlett-Packard Development Company, L.P.

[91] *TECHNICAL NOTE: Low Power Function of Mobile RAM? Auto Temperature Compensated Self Refresh (ATCSR)*. 2005. Elpida Memory, Inc.

[92] *DATA SHEET: DDR3L SDRAM 240Pin Unbuffered DIMM Based on Micron 4Gb E-die*. 2013. Ramaxel Technology Co., Ltd.

[93] Restle, P.J., J.W. Park, and B.F. Lloyd, *DRAM variable retention time*. in *Electron Devices Meeting, 1992. IEDM '92. Technical Digest., International*. 1992.

[94] Chan-Kyung, K., et al., *CMOS temperature sensor with ring oscillator for mobile DRAM self-refresh control.* in *Circuits and Systems, 2008. ISCAS 2008. IEEE International Symposium on.* 2008.

[95] *DDR3 SDRAM Specification.* 2010. Samsung.

[96] *Low Power Function of Mobile RAM? Auto Temperature Compensated Self Refresh (ATCSR).* 2005. Elpida Memory, Inc.

[97] *Mobile DRAM Power-Saving Features and Power Calculations.* 2005. Micron Inc.

[98] *Application Notes: TCSR-How to reduce self refresh current according to temperature changes.* 2002. Samsung.

[99] *Octopus 8-Port DRAM for Die-Stack Applications*, http://www.tezzaron.com/memory/Octopus.html. 2010. Tezzaron, Inc.

[100] Kang, U., et al., *Co-architecting controllers and DRAM to enhance DRAM process scaling.* in *The Memory Forum.* 2014.

[101] Lin, J., et al., Software thermal management of dram memory for multicore systems. ACM SIGMETRICS Performance Evaluation Review, 2008. **36**(1): pp. 337–348.

[102] Lin, J., et al., *Thermal modeling and management of DRAM memory systems.* **Vol. 35**. 2007: ACM.

[103] Gurumurthi, S., A. Sivasubramaniam, and V.K. Natarajan, *Disk drive roadmap from the thermal perspective: A case for dynamic thermal management.* **Vol. 33**. 2005: IEEE Computer Society.

[104] Youngjae, K., S. Gurumurthi, and S. Anand, *Understanding the performance-temperature interactions in disk I/O of server workloads.* in *High-Performance Computer Architecture, 2006. 12th International Symposium on.* 2006.

[105] Skadron, K., T. Abdelzaher, and M.R. Stan, *Control-theoretic techniques and thermal-RC modeling for accurate and localized dynamic thermal management.* in *High-Performance Computer Architecture, 2002. Proceedings. 8th International Symposium on.* 2002.

[106] Zanini, F., D. Atienza, and G. De Micheli, *A control theory approach for thermal balancing of MPSoC.* in *Design Automation Conference, 2009. ASP-DAC 2009. Asia and South Pacific.* 2009. IEEE.

[107] Yang, G. and Q. Qinru, *Dynamic thermal management for multimedia applications using machine learning.* in *Design Automation Conference (DAC), 2011 48th ACM/EDAC/IEEE.* 2011.

[108] Bartolini, A., et al., Thermal and energy management of high-performance multicores: Distributed and self-calibrating model-predictive controller. Parallel and Distributed Systems, IEEE Transactions on, 2013. **24**(1): pp. 170–183.

[109] Zanini, F., et al., *Multicore thermal management with model predictive control.* in *Circuit Theory and Design, 2009. ECCTD 2009. European Conference on.* 2009.

Chapter 4
Active cooling

On-chip temperature sensors are essential components of DTM systems. Performance throttling through voltage and frequency scaling has the most immediate impact on power density and thermal response. However, at times, throttling alone is insufficient to bring a hot system back to a safe operating point. Furthermore, even if it were feasible to achieve a safe operating temperature via throttling, the price paid in lost performance may be too steep to endure. Therefore, additional intervention through active cooling solutions is required, particularly for high-performance systems to be able to operate these systems with high utilization. Furthermore, cooling is critical during pre-deployment functional testing and burn-in, chips are subject to above average-case thermal stress. In order for the functional test results to be reliable, the chips should not experience transient timing violations due to extreme thermal stress.

Active cooling generally refers to all mechanisms that require additional power input to force heat away from a given system towards the ambient air. They come with an energy cost associated with their operation, for example, motors to rotate fans, pumps and chilling units for liquid cooling, and input current used to bias a TEC. They direct the heat flow in a specific direction away from ICs and towards the heatsink and the ambient air. In the case of thermoelectric cooling, with a given input power spent on the device, a negative effective thermal resistance can actually be achieved resulting in the reversal of the thermal gradient. On-chip temperature sensors can also be involved in feedback loops governing the control of various active cooling solutions. For example, many server-type processor chips utilize temperature sensor readings to decide when to ramp up and down the cooling fan speed.

A simple yet widely used metric of the required or desired cooling ability of an active cooling system is the effective thermal resistance between the die's surface and the ambient air [1]. An approximation to the desired maximum thermal resistance to relieve heat produced on a die subject to TDP power consumption is shown in (4.1).

$$R_{eff}^{therm} = \frac{(T_{die} - T_{ambient})}{P_{TDP}} \tag{4.1}$$

This relationship defines the minimum expected thermal conduction quality such that the die temperature is not to exceed T_{die} when the system is consuming the maximum amount of power allocated by the TDP specification, at a given ambient temperature. It is understood that the system can have very short periods of activity exceeding this TDP level and correspondingly exceed the temperature limit

T_{die} up to another safety limit. Yet, those intervals should strictly not to be representative of the workloads, but rather only exceptions. Several components take part in this environment including all passive layers of heat conduction paths in physical contact with the die and all active cooling mechanisms that surround this inner layer. The effective thermal resistance can also be defined hierarchically between the die and the chip package and also between the chip package and the ambient. In the former case, the thermal resistivity is directly related to heat conductance of a physical solid object (or rather combination of several objects), including all materials layered between the die and the package casing. In the latter case, the thermal resistivity can be an abstract association with the ability of the active cooling method of choice, for example, cooling air creating convection paths within the interiors of the heat pipes, case, chassis, rack, or cabinet (depending on the specific target product and deployment method for the processor chip).

With the increasing dominance of localized hotspots in modern processor chips, a new metric has been introduced to differentiate between a chip with uniform power density across its area versus one where local concentration of power and heat is more prominent [2]. This metric aims to capture the ability of a cooling solution to relieve heat accumulated per unit area and its efficiency in dealing with non-uniform power densities. This new metric is referred to as the junction to case density factor (DF_{jc}) and it is defined as:

$$DF_{jc} \; \frac{R_{jc}^{eff}}{G_{jc}^{norm}}, \tag{4.2}$$

where G_{jc}^{norm} is the junction to case thermal resistance normalized to die area under uniform power density.

In the following, we will review major active cooling mechanisms and the types of support they receive from sensors. The discussion will start with the most common and perhaps least sophisticated of all active cooling techniques, which is air cooling. Liquid cooling has found commercial use in select processor chips used in HPC systems, for example, supercomputers. Finally, the emerging field of thermoelectric cooling will be discussed.

4.1 Air cooling

Forced air cooling has been the first and most widely used active cooling method. Cooling fans are attached in various shapes at different locations within the system. Some heatsinks are integrated with a cooling fan as depicted in Figure 4.1. Placing a fan on top of a heatsink to allow vertical flow of cooling air towards the heatsink is also referred to as impingement cooling. There are two types of implementations combining fans and heatsinks. Cooling fans can be attached directly on top of a heatsink as shown in the figure. Also, for more compact form factor and specifically for reducing the height of the structure, the heatsink fins are implemented as a pin array and the fan is embedded inside the array lowered into a cavity at the center of the array.

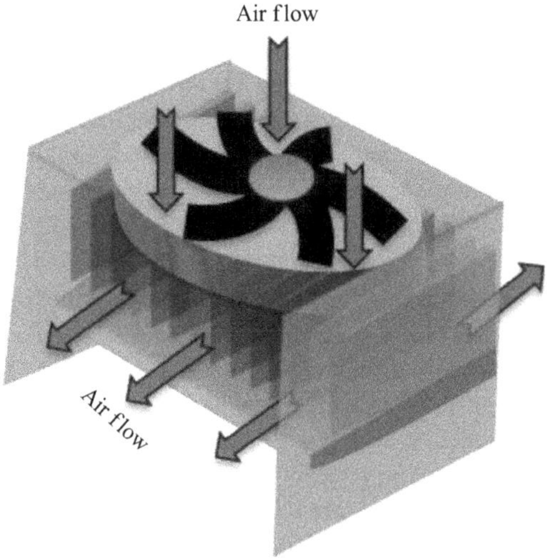

Figure 4.1 Heatsink with an integrated cooling fan. Air flows downward first from the fan and then through the fins of the heatsink outwards

In addition to fans combined with heatsinks, cooling fans are often attached within the case on the ventilation holes for pulling-in cool air and pushing-out warm air. Inside cabinets, rack mounted cooling fans are also placed to provide forced air through the boards and power regulators [3]. Placement of cooling fans is optimized after careful analysis of the airflow and resulting cooling within a given system. Several additional concerns of form factor, connectivity, height of components, interfacing, etc. influence the final layout and the final achieved cooling efficiency.

Another extension to heatsink structures to improve the effectiveness of forced air cooling is the addition of heat pipes with a larger remote heatsink, also referred to as a remote heat exchanger [4]. This structure essentially connects a local cooling block attached to the chip with another heatsink block placed farther away on the board. Pipes running through the block attached to the chip carry heat towards the remote heat exchanger. The pipe contains a fluid (it is usually water for processor cooling) and a wick (e.g., groove structures engraved into the pipe or powder materials). The fluid converts to vapor at the hot end of the pipe and gets attracted back to the heat exchanger via the wick for re-condensation. Heat transported towards the heat exchanger is removed with the cooling air forced through the fins of the remote heatsink. The effective thermal conductance of the heatpipe surpasses that of copper. These structures have been used for cooling laptop computers, such as those using Intel Celeron M processors, as well as in systems using processor such as Core i5, i7, and Xeon.

Cooling fans are defined with a few specifications in practice, including the dimensions (radius), rotations/revolutions per minute (RPM), the airflow provided, and the input power they require to operate. RPM for a cooling fan varies in relation to the specific type of fan used in different locations of the system.

Smaller cooling fans combined with the heatsinks range in size between 30 and 60 mm. Larger fans attached at the board or cabinet level range between 40 and 360 mm in diameter depending on the cooling requirement and available space. The speed of rotation has a direct impact on the cooling power. RPM used in cooling fans for processor chips can range between 1,000 and 4,000, while cooling fans dedicated for other components (e.g., hard disk, graphics cards) and fans used at system level might have higher RPM. The static pressure (SP) of air provided by a cooling fan and the airflow achieved by the fan help to define the effectiveness of cooling. SP of a fan signifies the impact of the fan on the airflow path against the resistance and obstacles. Hence, a fan with higher SP rating can force the same amount of air volume through tighter spacing and/or obstacles. The SP is therefore a more direct measure of cooling effectiveness for restricted spaces, such as fans attached to heatsinks. Hence, fans that are capable of delivering higher SP would be preferred in these situations. The pressure in air cooling systems is often measured in units of millimeter (or inch) of water (mm H_2O), which is the pressure exerted on the base of a column of 1 mm of water at 4°C. Fans that deliver high airflow are generally better suited for unobstructed and less constrained spaces such as rack mounted fans and cabinet fans. Airflow is defined as the air volume delivered per unit time, often using the unit of cubic feet per minute (CFM).

For a given system with a specific cooling need and known air volume, a fan is configured to provide airflow for one specific pressure level. For instance, the flow impedance of various models of many-core Intel Xeon co-processor boards are rated between 0.14 in H_2O and 0.406 in H_2O. Figure 4.2 illustrates the relationship between the capabilities of a given fan and the requirements of the system that needs to be cooled. Any given fan can deliver a specific amount of airflow in presence of a given

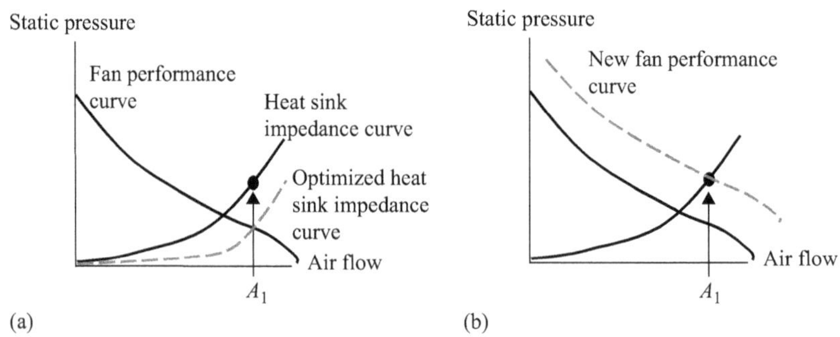

Figure 4.2 Determining cooling fan specifications considering the relationship between the performance of a given fan subject to a system's resistance to airflow and the minimum amount of airflow required by the same system. (a) If the airflow requirement cannot be met by the fan operating with a given SP and resulting total airflow, then one option is to improve (reduce) the impedance of the system, lowering the impedance curve of the heatsink. (b) Another option is to deploy a higher performance fan, raising the performance curve of the fan

pressure dictated by the system's impedance. Figure 4.2(a) depicts the impedance of a heat sink under a range of airflow levels. It also depicts the performance curve of a given fan that can deliver specific amounts of airflow into a given pressure. Let us assume that the heatsink's thermal design requirement dictates an amount of airflow corresponding to A_1. The fan depicted in Figure 4.2(a) does not meet this requirement. An option to mitigate this is to improve the impedance of the heatsink, which is illustrated with a dashed curve in this figure. This may involve using higher efficiency material for the heatsink, changing the geometry, etc. The other option is depicted in Figure 4.2(b) where the fan is replaced with a higher performance fan that is capable of delivering A_1 airflow for the original impedance of the heatsink. This may involve using a larger fan, higher input power to the fan, etc.

The relationship between necessary airflow for given cooling requirements can be approximated using the heat transfer equation [5]:

$$Q = C_p^{air} W \Delta T, \tag{4.3}$$

where Q is the amount of heat transferred, $C_p^{air}$is the specific heat of air, W is the mass of airflow, and ΔT is the temperature differential between the inlet point and exhaust of the air flow path. The mass of air transported can be calculated as:

$$W = Airflow \times Density \tag{4.4}$$

Re-arranging these relationships to calculate airflow for a given amount of heat to be removed with a corresponding temperature gradient yields an approximation as follows:

$$Airflow = \alpha \frac{Power}{\Delta T} \tag{4.5}$$

The coefficient α is set to 3.2 if temperature is measured in units of Fahrenheit or 1.78 is the unit of temperature in degrees Celsius [6]. The power consumption of the system to be cooled directly relates to its heat production, hence, Q is represented with the power dissipated in the system.

It has been reported that high density enclosures hosting blade servers would require 2,000 CFM to cool 15–25 kW for one particular Dell system [6] and 2,500 CFM to cool another system of six 7U blade server frames dissipating 18 kW of power [7].

The relationship between the fan speed and the fan performance metrics of pressure and airflow can be expressed using (4.6) and (4.7). There is a linear relationship with the rate of increase of fan speed and airflow. The relationship between SP and fan speed is quadratic.

$$Airflow_2 = \frac{RPM_2}{RPM_1} Airflow_1 \tag{4.6}$$

$$SP_2 = \left(\frac{RPM_2}{RPM_1}\right)^2 SP_1 \tag{4.7}$$

Power consumption of the cooling fan itself is directly related to the fan speed, which in turn is dictated by the required airflow rate. The input power supplied to a cooling fan can be calculated from its specifications as the product of the operating voltage and amount of current drawn by the fan's motor at a given rotation speed:

$$P_{in}^{RPM} = V_{fan} I_{fan} \tag{4.8}$$

The output power of a fan is expressed as a function of the pressure and airflow that it provides:

$$P_{out} = \frac{Pressure\,(inchH_2O) \times Airflow\,(CFM)}{8.5} \tag{4.9}$$

Finally, the efficiency of a cooling fan is defined as the ratio of its output power to the input power.

$$\mu_{fan} = \frac{P_{in}}{P_{out}} \tag{4.10}$$

For a fan with a given efficiency in its specifications, the power consumption of the fan can now be calculated with the product of its efficiency and output power that it delivers. Since SP and airflow (the two components of output power) have a quadratic and linear dependence on fan speed, respectively, the power consumption of a fan can be approximated as a cubic function of its speed. The power consumption of cooling fans in high-end servers can reach 40 W for 1U racks [8] and exceed 200 W in 2U racks [9]. The contribution of the cooling fans (including processor, memory chip, and I/O component fans) to the total system power consumption can rival that of the processors and reach up to 51% [10].

In addition to the purely performance-related specifications of a cooling fan, there is one more important parameter concerning the acoustic noise generated by the fan. Acoustic noise impacts the user's experience, especially for desktop and laptop computers. While discomfort due to noise is subjective, the relative effect of fan operating conditions on acoustic noise can be quantified using the decibel (dB) metric. The fan speed is the most important factor in acoustic noise and it can be expressed through the following relationship [11]:

$$Noise_1\,(dB) = Noise_2\,(dB) + 50 \times log_{10}\left(\frac{RPM_1}{RPM_2}\right) \tag{4.11}$$

Increasing the speed by 50% would correspond to an increase of noise by about 8.8 dB. Another study has shown that the acoustic noise level increases by about 10 dB as air flow rate increases by about 52% [12]. Recalling that airflow is linearly related to fan speed, this data is consistent with the approximation of (4.11).

Overall, forced air cooling systems are optimized as part of a holistic thermal design methodology, where (i) thermal impedance of the underlying system is minimized to the extent possible under form factor and placement constraints; (ii) cooling fan with the best fitting performance and efficiency for a given impedance behavior and airflow requirement is chosen; and finally (iii) the cooling fans are managed dynamically to operate at the lowest speed possible for as long as possible to minimize acoustic noise and more importantly the fan's power consumption. Clearly, meeting this objective also minimizes acoustic noise. A variety of control mechanisms exist to manage the fan speed of heatsink, case, and cabinet fans for this purpose. In the following, we will review these management schemes.

4.1.1 Managing cooling fans

In recognition of the impact of cooling fans to total system power consumption, a number of studies focused on managing fan speed efficiently to adapt to varying cooling demands of a system's heatsink.

One of these approaches formulates the effective thermal resistance of a heatsink as a function of fan speed (increasing fan speed decreases effective thermal resistance of the heatsink) and manages this as a variable parameter while maintaining the heatsink temperature at a desired equilibrium [13]. A control algorithm is then proposed to minimize total power, which is composed of dynamic power of processors, temperature-dependent leakage power of processors, and fan power. This is achieved by co-optimizing the fan speed and voltage and frequency scaling factor dynamically. A control theoretic method is adopted, where a feedback loop is enabled through real-time temperature measurements. Each time the temperature sensors report overheating, an optimization problem is solved to determine energy-optimal (fan speed, voltage/frequency) setting and this setting is maintained until the present task is complete or the sensors report another violation of the safe thermal limit.

A multi-input multi-output fan controller was proposed for blade servers [14]. The optimization problem is formulated with the objective of minimizing total fan power over all fans attached to the server enclosure, while the maximum junction temperature reported within each blade is kept under a threshold. In addition, fan speed is constrained with an upper and a lower bound. This optimization problem is solved dynamically, adjusting to the time varying nature of workloads. The cubic relationship between fan speed and power is captured with a third-order polynomial. In order to attack this inherently non-linear optimization problem, it is divided into two sub-problems. The first step identifies the minimum cooling requested by each individual blade locally to maintain its junction temperature under the given threshold. Minimum required airflow rate for any blade is capped at the maximum available airflow rate of the fans. Once, the minimum required airflow rate is determined for each blade, the global problem is solved in the second stage, where a binary search algorithm is employed to iterate through the ranges define by the minimum required airflow and maximum available airflow.

A fan controller that uses continuous monitoring of processor and fan power to adapt fan speed for energy optimization was proposed [15]. Extensive characterization of a server system was used to collect separate processor and fan power and corresponding processor temperatures at varying levels of system utilization. Special benchmarks were used to stress the system at different levels. Power, temperature, and fan speed data collected from these characterization runs are then used to perform model fitting to generate models for the active and leakage power components as a function of utilization and temperature. As different fan speed settings are scanned during the characterization runs, for each utilization level, the optimum fan speed setting that minimized the total energy consumption is also identified. Using these fitted models a LUT is generated indicating the energy optimal fan speed choice at different utilization levels. At run-time, utilization is polled frequently (every second) and the LUT controller determines the appropriate fan speed. The choice of relying on utilization information rather than direct temperature measurements is justified with the argument that trends in utilization will be early precursors of imminent thermal events and the system will be able to react proactively. On the other hand, it is possible for the controller to elevate fan speed to soon in response to transient and short lived utilization peaks, potentially running the fan faster for longer than needed.

A recent study returns back to the paradigm of utilizing online temperature sensor measurements for control. However, the imperfections introduced due to the latency in obtaining a sensor's reading and quantization errors can cause stability concerns for the controllers relying on these sensors. This particular study focuses on the stability of dynamic fan speed control schemes manifested through fan speed oscillations [16]. In order to provide guarantees on stability a PID controller is introduced. The response function of the PID controller at step $k + 1$ is comprised of three terms each scaled with a gain factor. These are (i) the proportional gain scaled by the temperature measurement offset; (ii) integral gain for factoring the temperature observations up to step k; and (iii) derivative gain factoring the difference of temperature observation error between the most recent two observations. The system is first observed in a state of indefinite oscillations at steady state. The system is placed in this state by tuning the proportional gain parameter only. The resulting period of oscillation is measured and the global gain coefficient is identified. Using these two observed values, the three individual gain parameters are calculated using the Ziegler–Nichols closed loop tuning method [17].

Another PID controller for fan speeds in high density servers aims to achieve self-tuning [18]. A PID neural network is utilized to tune gain parameters of the PID online. Similar to other previous methods, a third-order polynomial is used to fit the fan power consumption into a function of fan speed. The objective of the controller is to minimize fan power consumption during a transient period between time instances t_1 and t_2. Unlike the previous PID method, which used a static parameter tuning scheme, this method uses online learning leveraging a neural network. In fact the three-layer neural network contains the actual PID controller within its intermediate layer.

A learning-based method emphasized the relationship between fan speed and leakage power [19]. Since fan speed influences temperature, which in turn affects leakage power, there exists an optimal point where the combined fan and leakage power reaches its minimum. This fan speed control method uses hill climbing to train the controller for optimal fan speed during the early phase of a given workload. During this phase the workload is assumed to be stable and representative. This method is suitable for high-performance computing applications, where long running jobs dominate the workload. After the learning phase, the fan speed determined for this phase is adopted until a significant change in the workload is detected. The search space for optimal fan speed is assumed to lie in a convex space. The hill climbing method basically performs a search in this space by starting with an arbitrary initial solution and exploring the neighborhood by re-positioning the central point of the search each time a better solution is found. The time required for the search is constrained by restricting the neighborhood in various ways.

A real-time fan controller for servers employs a reduced thermal model to predict the impact of a given air flow mass on the temperature of servers [20]. The reduced model relating a target temperature and air flow mass for a given heat output is derived using the proper orthogonal decomposition, which is a model order reduction method. The resulting controller takes the observed system temperature and processor power measurement as input and applies them to the compact model to derive the mass flow rate. For an initial estimation of the mass flow rate, the corresponding temperature is predicted. If the predicted temperature is within a pre-defined proximity to the reference temperature set by the system's manufacturer, the mass flow rate is converted to a corresponding voltage value to drive the fan. If the predicted temperature is much lower than the reference temperature, the mass flow rate is decreased by a small amount and the temperature prediction is repeated. Similarly, if the predicted temperature is higher than the reference temperature by a larger difference, then the mass flow rate is increased by a small amount. This inner loop is repeated for a few iterations if needed.

Various aspects of optimization concerning the operation of a cooling fan have been considered by a study, including the elevated acoustic noise due to abrupt changes in fan speed, choices in controlling the supply voltage of the fan (regulators vs. pulse width modulation (PWM)), choices in monitoring fan speed at a fine granularity through monitoring the tachometer signal of the fan versus defining coarse grain speed steps, and establishing an optimal temperature point to start ramping up fan speed to prevent imminent thermal emergencies [21].

As we observe in many cooling fan optimization methods, co-optimization of computation power and fan power is a common thread. A method with this objective views fan power optimization as an integral part of DTM [22]. Again, another common observation drives this method concerning the convexity of the space defining the relationship between fan speed, fan power, temperature, and temperature-induced leakage power. The effective heatsink-to-ambient thermal resistance is the main parameter that is adjusted during thermal management to maintain the fixed thermal gradient under varying processor power according to the relation shown in 4.1. The effective resistance is modeled as a function of air velocity provided by the fan and a set of physical coefficients as follows:

$$R_{HA}^{eff} = \left(c_1 v_{air}\left(1 - e^{\frac{c_2 {v_{air}}^{c_3} + c_4}{c_1 v_{air}}}\right)\right)^{-1} \tag{4.12}$$

By substituting the cubic relationship between airflow rate and fan power, the effective thermal resistance can be expressed as a function of fan power directly.

Leveraging this formulation, the fan control method adjusts the fan power through PWM during thermal management while also making voltage and frequency scaling decisions for minimized total energy.

In addition to the large body of academic work on controlling cooling fan speed for energy optimizations under thermal constraints, commercial products also employ dynamic management for fans. Prominent mechanisms used in commercial products were discussed as part of the DTM schemes using on-chip temperature sensor feedback in Chapter 3. These schemes mainly operate by responding to a thermal emergency by boosting the fan speed to the maximum level until the junction temperature reported from within the processor cores drops below the safe threshold. This occurs in a coordinated fashion with dynamic voltage and frequency scaling. Once the maximum junction temperature is reported to be below the safe limit, fan speed is reduced to minimize fan power and acoustic noise. We refer the reader to Chapter 3 for more details on managing fan speed as part of the DTM mechanisms in commercial processors.

Finally, there are investigations of managing forced air cooling for other components beyond those that are located on the motherboard and extending outside of the server domain. One study was performed on forced air cooling of the battery component in a portable system [23]. Drastic changes in battery temperature observed during continuous charging and discharging cycles motivated this study. An interesting aspect of this scenario is that power for the fan cooling the battery is supplied by the same battery. When the fan is active it will impact the charging capacity of the battery. At the same time, the fan helps to reduce the battery temperature improving its lifetime. This is a trade-off that requires careful exploration. A reinforcement learning-based technique is employed within a hierarchical framework to first determine the improvement in state of health of the battery for a given amount of cooling investment. At the higher level, a dynamic programming-based algorithm is used to determine the best trade-off between cooling level (i.e., determining whether the fan should be off or if it should be on, at which speed setting) and amount of work supplied within the lifetime of the battery. The end of a battery's lifetime is defined as the point in time where the reliability of the battery drops to 80% of its initial state of health.

4.1.2 Beyond fan-based forced air cooling systems

While mounted cooling fans and impingement coolers integrated with heatsinks are used in practice, there are other developing technologies, which attracted attention, yet they are still at the prototype and evaluation stage. One example is the Air-bearing Heatsink (also called the Sandia Cooler) developed by the Sandia National Laboratories [24]. Essentially, this design merges the concepts of heatsink and fan into one structure. Aluminum cooling fins are shaped as rotating blades and suspended across an extremely thin air gap (~0.03 mm) above the heat pipe baseplate. A brushless motor located at the center propels the blades.

Microjet (also referred to as jet impingement) technology has also been investigated for microelectronics cooling [25–29]. This cooling technology finds

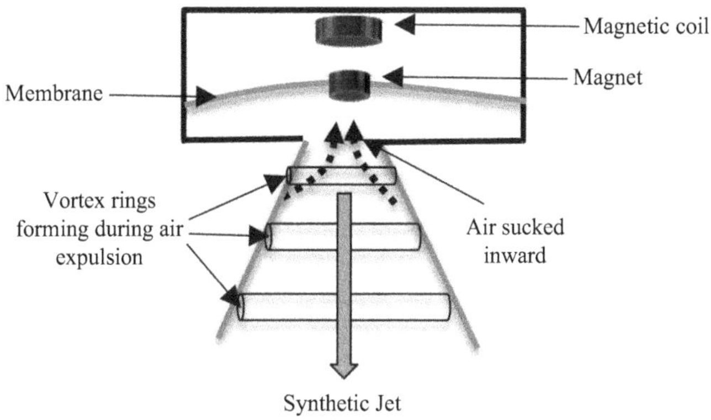

Figure 4.3 Illustration of a cavity device for generating synthetic jet cooling

practical use in systems with excessive heat flux, such as equipment in manufacturing industries, lasers, and light emitting diodes. Early proposals involved creating impingement jets using pumps ejecting the jets through miniaturized cavities targeting the hot surface. More recent approaches involve creation of synthetic jets [30–33]. Synthetic jets are formed by a device comprised of an air cavity that is bounded by a flexible membrane on the top side and an orifice on the bottom side looking pointed at the hot surface [25]. This device is embedded in fluid as opposed to delivering fluid through specialized plumbing and channels. The vibration of the membrane draws air into the cavity and then ejects it through the orifice. The entrainment of air is followed by vortex rings formed during ejection, which expulse the surrounding fluid in the form of a synthetic jet. The vibration of the membrane can be created by placing a magnet on the membrane and applying an AC signal to coils placed at the back of the membrane to create alternating magnetic fields. Recent investigations focused on modeling these devices, deriving the effective thermal resistance of a heatsink in present of synthetic jet cooling, and sensitivity studies on device parameters (such as shape of the orifice, excitation frequency) [34]. A simple illustration of this device is depicted in Figure 4.3.

Piezoelectric fans are another technology promising frictionless movement of a cantilever blade driven by a piezoelectric transducer [35–40]. A strip of piezoelectric material (e.g., PVF_2) is excited with AC voltage causing the material to expand and contract. This flexing movement is amplified with a cantilever attached at the tip of the piezoelectric strip. Airflow is created moving away from the cantilever's tip. The cantilever can be placed facing a heatsink sideways to push the airflow through the heatsink fins. These types of fans are promising as lower noise and lower power fan systems and further investigation for practical use is needed. Some investigations focus on the applicability of this technology for power constrained devices such as laptops and mobile phones [40, 41].

Another concept of frictionless fan blades combines the idea of cantilever type blades of piezoelectric fans and the magnetic operation of synthetic jets [42–44].

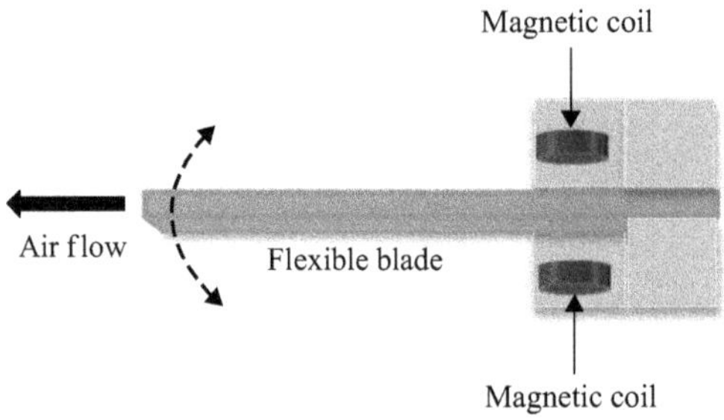

Figure 4.4 Illustration of a frictionless blade for forced air cooling driven by magnetic fields

A flexible polymer-based or steel blade is moved using a magnetic field to create airflow in the perpendicular direction as illustrated in Figure 4.4.

Finally, an intriguing cooling technology uses electric charge to ionize air molecules to achieve heat transport [45–48]. This technology, like the piezoelectric and magnetic blades, is promising particularly for small form factor devices such as laptops and mobile devices. The cooling system utilizes two cathodes to create an electric field. Electrons drifting towards the positive cathode knock off electrons of oxygen and nitrogen molecules in the air ionizing them. As the ions move towards the negative cathode they create the "ionic breeze."

4.2 Liquid cooling

The evolution of chip cooling has started out with passive heatsinks and moved on to forced air cooling. While several advancements to air cooling with potentially promising power, form factor, low noise, and high efficiency are still under investigation as discussed above, cooling technologies have also developed along other avenues concurrently. The technology that has found commercial success and use in practice is liquid cooling. Liquid cooling has a long history starting with IBM's introduction of water cooling for removing heat from stack of boards. At the present time, liquid cooling is used in practice to mitigate heat in multiple layers of the stack from racks to individual chips. This technology incurs new design and manufacturing overheads since the cooling systems involve liquid circulation with the associated additional components (pump, tubes, microchannels fabricated in heatsinks) and need to ensure proper isolation of the coolant. The higher design costs are feasible for systems with extreme thermal stress that are generally encountered in high-performance computing. Datacenters and supercomputers utilize liquid cooling at the rack level. One of the prominent examples of liquid cooling at the chip level is the IBM Blue Gene/Q chip used in supercomputers. It hosts 18 PowerPC cores and uses liquid cooling embedded in its heatsink. Another domain that is attracting interest for liquid cooling is 3D ICs with a large body of academic studies on various design aspects. Finally, there is interest from the

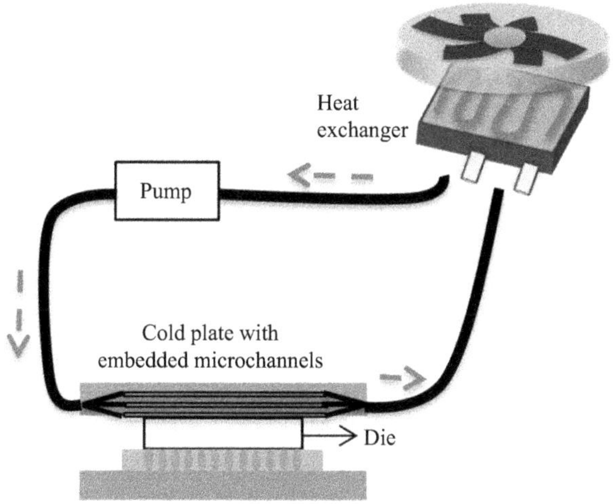

Figure 4.5 Illustration of a microchannel liquid cooling system utilizing a remote heat exchanger to cool a microchip

overclocking community in retrofitting computers with liquid cooling [49]. A demonstration of adding liquid cooling to a Mac Pro revealed significant thermal headroom to overclock the computer from 2.8 to 3.16 GHz [50].

Regardless of the scale at which they are applied, liquid cooling systems have two main components. The first is made up of a pump to circulate the liquid and channels integrated with a cold plate (heatsink) to carry the flow and enable heat removal from the hot surface that they come into contact with. The second component is the heat exchanger, which is located remotely, connected with the heatsink through tubes containing the liquid. The heat exchanger receives the warm liquid and heat is extracted from the liquid by the system fans blowing cold air over the exchanger. Once heat is released to the ambient, the liquid can be re-circulated by the pump. Figure 4.5 illustrates the major components of a single-phase liquid cooling system embedded into the heatsink. In this figure, liquid cooling for an IC is depicted. The system's major components are a closed loop circulation system with a coolant, a pump to direct the coolant, a cold plate engraved with microchannels, and a heat exchanger consisting of a heatsink cooled with a fan. Liquid cooling systems deployed in datacenters and supercomputers follow the same basic principle at the macroscale. In these large systems, heat exchangers are often not cooled by fans. Instead, they use liquid-to-liquid heat exchangers. The warm coolant liquid returning from the computing systems passes through a chiller to remove heat. While this provides stricter guarantees on operating temperatures for these high-performance systems, it results in another cost item on the energy bill spent for cooling. In response to this, chiller-less data centers that only utilize outside air temperature to remove heat from warmer coolant liquid have been proposed [51, 52]. Also, supercomputers that use warmer coolants have been deployed in recent years [53, 54].

Liquid cooling applied at the chip level involves microchannels (miniature grooves) that are embedded within the heatsink. A coolant liquid (usually water) is circulated through these channels laterally across the hot surface of the heatsink. An early proposal for a heatsink housing microchannels argued for scaling the liquid cooling heat exchanger technology at the time to microscales so that it can be applicable to ICs [55]. Since then a large body of work has resulted from efforts to design and manufacture nanoscale cooling channels [56, 57], alternative coolant liquids [58, 59], integration of the channels and liquid pumps with the chip packages [60], management of liquid pumps for optimal flow and pressure, and design of the heat exchanger that collects the coolant after it travels through the heatsink [61–65].

Microchannel liquid cooling systems can be broadly categorized as the single- and multi-phase systems. Single-phase systems maintain the coolant only in liquid form throughout the cooling process, whereas multi-phase systems allow the coolant to transition from liquid to vapor and vice versa. The flow of the coolant liquid within channels can also be broadly categorized in one of two types: laminar flow and turbulent flow. Laminar flow is the organized movement of liquid in a steady state through a channel. This results in even layers of liquid molecules moving at fixed velocities along their respective layers. The turbulent flow disrupts the even separation of layers and liquid molecules no longer remain within a single layer through the flow path, changing velocity and placement throughout the path. Heat is transferred from the heat source (i.e., chip) to the liquid in the channel through radiation and convection through the walls of the channels and by conduction from the wall surface to the fluid inside. The effective thermal resistance (or heat transfer coefficient) is used as a metric for the cooling efficiency just like it has been done for forced air cooling systems. The heat transfer is then governed by the following relationship,

$$Q = U A\Delta T, \tag{4.13}$$

where Q is the amount of heat transferred, U is the heat transfer coefficient, A is the surface area, and ΔT is the temperature gradient between the inlet and the outlet. The convective heat transfer between the channel boundary and the coolant liquid can also be defined with the dimensionless Nusselt number, which denotes the ratio between the convective and conductive heat transfer at the boundary:

$$Nu = \frac{hD}{k}, \tag{4.14}$$

where h is the convective heat transfer coefficient, D is the characteristic dimension of the channel (i.e., diameter of a circular channel), and k is the thermal conductivity of the liquid. For laminar flow in a channel, this quantity is a constant yielding an inverse relationship between the heat transfer coefficient and channel dimension.

Another dimensionless parameter used to define liquid cooling systems is the Reynold's number Re, which helps to formulate the flow pattern of fluids. Laminar flow is associated with a low Re number, while turbulent flow is characterized by a high *Re* number. This parameter is used to model and characterize pressure drop in microchannels.

There are two more dimensionless numbers used for expressing fluid heat transfer mechanisms: the Rayleigh number *Ra*, indicates a relationship between buoyancy and viscosity of a fluid and its thermal diffusivity, the Prandtl number *Pr*, also describes a relationship between kinetic properties and thermal diffusivity of fluids.

The Nusselt number can also be expressed as a function of *Ra* and *Pr* as follows [66]:

$$Nu = \left(0.825 + \frac{0.387 Ra^{\frac{1}{6}}}{\left[1 + \left(\frac{0.492}{Pr} \right)^{\frac{9}{16}} \right]^{\frac{8}{27}}} \right)^2 \tag{4.15}$$

Another important metric for liquid cooling concerns the liquid pressure within the microchannels. The pressure drop along the cooling path is particularly important as it rests on a balance against heat transfer efficiency of the microchannel. The heat removal correlates with mass flow rate. On the other hand, with increasing mass flow rate the pressure drop along the channel will decrease. Either the pressure drop must be compensated with additional pumping power or degradation in efficiency is to be suffered. The pressure metric is often expressed in terms of pressure drop per unit [67]:

$$\frac{\Delta P}{L} = \frac{f \rho V^2}{2D} \tag{4.16}$$

ΔP is the pressure drop along the channel, L is the length of the flow path, V is the mass flux, f is the friction factor, D is the characteristic channel dimension, and ρ is the fluid density. f is referred to as the Fanning factor describing the ratio of the wall shear stress to the kinetic energy per unit volume [68]. The product of the Fanning factor and the Reynold's number is a constant for laminar flow for a given channel geometry and in the turbulent flow scenario these two parameters follow a correlation as well.

In the presence of the fundamental heat transfer relations described above, geometric parameters, materials, the cooling requirements of the chip served by the cooling system, and the available energy budget dedicated for the cooling system all factor into determining the type of flow present in the system and the resulting heat transfer efficiency of the liquid cooling system. These parameters can be optimized according to cost-efficiency trade-offs. In the following, these parameters will be reviewed.

4.2.1 Optimization of efficiency and cost of liquid cooling systems

Microchannel geometry: The dimensions and placement of the microchannel structures have a direct relationship with the cooling efficiency. The channels can be circular with the diameter defining the characteristic dimension or non-circular (e.g., rectangular), where the shorter side of the cross-sectional area can be considered as the defining parameter of geometry. Particularly, for the laminar flow, the constant Nusselt number dictates that the convective heat transfer coefficient will increase by several orders of magnitude as the channel dimension enters the nanoscale realm. On the other hand, extremely small channel dimensions exacerbate pressure drop and also the surface area available for conduction will be less. Therefore, in order to achieve a certain amount of flow rate the pressure drop induced by the channel's geometry needs to be compensated for.

Trade-off can be pursued along multiple dimensions of the design space. For instance, a study has compared two types of cold plates attached to processor chips: a relatively high performance, high-cost copper cold plate with channels machined as grooves into the structure and a relatively low performance, low cost copper tube embedded in an aluminum cold plate [69]. All cold plates under consideration use rectangular channels with water as coolant. Varying widths for channels of both cold plate types have been analyzed. The machined microchannels were able to achieve high heat transfer at relatively lower flow rates despite experiences higher pressure drop. The less expensive embedded tubes require higher flow rates with lower pressure drop.

In addition to the dimensions, other design options are flow direction and channel density. A number of channels can be placed in parallel with a single link of flow between the inlet and the outlet. The number of parallel channels per unit area of the cold plate can be a design parameter. The width of the channels and the total surface area of the cold plate, as well as the acceptable design complexity collectively determine the feasible density that can be created. Another alternative is to create a serpentine layout with the channels using multiple serially linked segments. Each segment may contain multiple parallel channels as well. These options are depicted in Figure 4.6.

The ratios of the width of the microchannel to the depth and the width of the fin (width of the borders separating individual channels) to the depth can be used as design parameters to optimize. A study formulated this problem with an associated constraint on the total pumping power [70]. The objective function is defined as the effective thermal resistance, which needs to be minimized. The average flow velocity is expressed as a function of pumping power and geometric features of the channel, Reynolds number, Fanning friction coefficient, and physical coefficients.

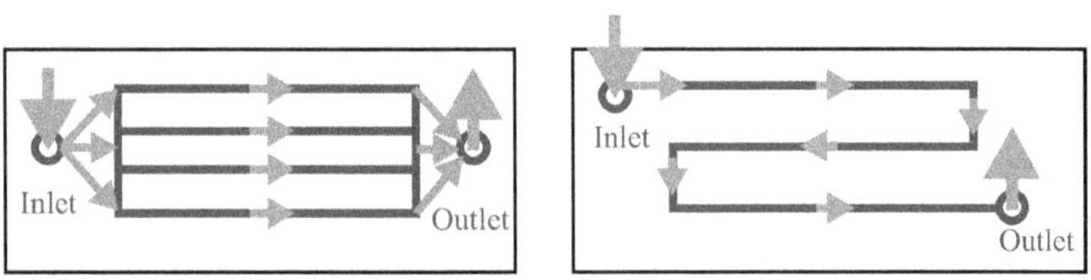

Figure 4.6 Alternative configurations for the liquid cooling channels. The illustration on the left depicts parallel channels of flow, while on the right, a serpentine topology is depicted

The pumping power was constrained by a given range that is consistent with the state-of-the-art micro pumps (0.01–0.8 W) and the search space for the dimension variables is explored by comparing the resulting flow velocity to conform to the pumping power range. The aspect ratios (channel width to depth and fin width to depth) were constrained within the ranges of (0.1–0.25) and (0.04–1), respectively. One of the main conclusions was that the thermal resistance is more sensitive to channel width to depth ratio than fin width to depth ratio around the optimal point. Different cross-sectional shapes other than circular and rectangular can also be considered. One study investigated trapezoidal channels created by chemical etching with a sidewall angle [71]. Another method for customizing microchannel geometry for improved heat transfer is the concept of channel width modulation [72]. Starting from the basic inverse relation between channel width and the Nusselt number, the effective heat transfer coefficient experienced by the underlying chip is derived by projecting the heat transfer coefficient through the side wall surfaces onto the vertical heat transfer coefficient. The resulting relationship, shown in (4.17) the height (H_C) and width (W_C) of the channel and the total width (W) of the structure including the fins surrounding the channel.

$$h^{eff} = h\frac{2H_C + W_C}{W} \tag{4.17}$$

By allowing a channel structure with non-uniform channel width customized as a function of the length of the channel, the effective convective resistance of the cooling structure is modulated. As the channel width is shrinking towards the outlet, the increasing temperature gradient experienced along the fluid inside the channel is compensated for with an improved convection.

The choice of parallel versus serial layout of multiple channels has also been investigated from the point of uneven pressure distribution along the channels. One possible disadvantage of long running parallel channels has been identified as steep pressure variation between the two endpoints. A HY layout has been proposed to mitigate this problem, where segments of parallel channels are joined at an intermediate point resembling a ladder structure [73].

Another design option is to consider stacking of microchannels as shown in Figure 4.7. The thermal resistance of a single layer of parallel channels can be reduced by 30% by introducing a second stack of microchannels [74]. A follow up study on the optimization of such a structure identified the channel width, fin width, and the aspect ratio (width to length of channels) as the primary design parameters [75].

Design of microchannels can also be subject to two choices: a smooth interior surface of the channel versus a grooved structure within the channel to increase the surface area. The ratios of the groove depth to the microchannel height, the groove pitch to the microchannel height, the groove diameter to pitch, and the microchannel width to height are used as design variables [76]. A pareto optimal front among design options was generated and it was found that the ratio of the groove pitch to microchannel height is the most Pareto-sensitive variable. While channel width to depth was found to be most influential on thermal resistance for smooth channel surfaces [70], the introduction of grooves indeed creates a significantly

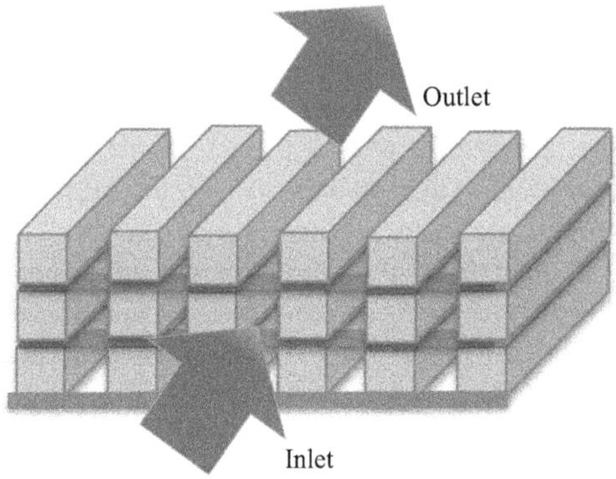

Figure 4.7 A multi-tier stack of microchannels

strong new design variable. A recent study also investigated a HY geometry combining grooved channel interior with smooth interiors along the channel's entire length [77]. Other geometries, such as T-shaped fins, posts placed within the channels, and staggered fins have also been explored [78].

Flow rate and micro pump design: Adjusting the flow rate for the cooling requirement and mitigating the pressure drop along microchannels are two main issues concerning the flow management. The water mass flow rate that is required to maintain a fixed outlet to inlet temperature difference under heat dissipation Q is expressed as follows:

$$\dot{m} = \frac{Q}{c_{p(T_o - T_i)}}, \tag{4.18}$$

where c_p is the specific heat of water. With increasing heat dissipation from the underlying device, higher mass flow rate is demanded from the cooling system. This system management challenge resembles dynamic performance throttling under a thermal constraint for processors. As the temperature increases performance needs to be throttled down. In the case of liquid cooling, flow rate must be scaled up. One option would be to determine a flow rate for the worst-case cooling need. This would naturally result in largest pumping power consumption. In response to this, a variable flow rate control mechanism has been proposed [79].

Besides, managing the flow rate of a cooling system, the design of the micro pumps serving the cooler is another active domain of research and development. There is a plethora of design approaches for micro pumps ranging from mechanical rotary pumps to electromagnetic diaphragms, piezoelectric actuators, shape memory alloy actuation, electrostatic actuation, electrowetting, and electroosmotic pumps [80, 81].

4.2.2 *Liquid cooling in 3D ICs*

With the advent of vertical integration, a new family of 3D ICs emerged with more aggressive cooling demands and unique thermal challenges. Majority recent liquid cooling studies had a 3D IC-centric focus. Therefore, it is appropriate to dedicate a separate discussion to the application of liquid cooling for 3D ICs. Early treatment of this subject concerned with liquid cooling applied to multi-chip modules [82, 83]. Recent investigations have introduced innovations for optimized cooling of multi-tier ICs.

One of the early studies utilized computational fluid dynamics (CFD) analysis to assess the thermal behavior of stacked 3D ICs and design an optimal thermoelectric-liquid HY active cooling system [84]. The liquid cooling system uses a water cooled heatsink attached to the heat rejection side of the TEC with a single segment channel that is wider at the inlet. The liquid is managed in turbulent flow regime with constant mass flow rate at the inlet and constant pressure at the outlet. CFD-based models of a 4-tier prototype chip with emulated heat sources and the HY cooling system was derived and its results were validated against measurements collected from the testbed system.

Another modeling effort aimed at better understanding of the thermal interactions between liquid cooling and a 3D IC and assess the peak performance that should be expected from liquid cooling [85]. The study established promising indications of success for 3D integration of liquid cooling, with heat densities up to 135 W/cm^2 removed by the cooler. This study modeled a fully integrated cooler, where channels are embedded between the tiers in the stack. Integration reveals new parameters for optimization. For instance, in a multi-channel cooler, flow rates and pressure along channels at different layers need not be identical. In fact, the heat removal needs of different locations would dictate differing cooling solutions. Furthermore, while different dynamics come into play within the interior of each channel, they interfere with each other, if the cooling system is built to share common inlets and exit manifolds and a common pump. An alternative is to deploy dedicated resources for each layer's channels. This study highlight the modeling challenges of these alternatives and simplifying assumptions that are needed to create a computationally feasible thermal model for the overall system. Models derived from heat transfer equations, considering channel geometries, fluid properties and the heat generation of the underlying IC have been developed for both single-phase and two-phase liquid cooling systems.

A composite multilayer stack of chips each utilizing a layer of microchannels with a pair of inlet/outlet ports was modeled with special consideration on the computational complexity of detailed CFD modeling [86]. In order to manage the complexity, the proposal is to represent the cooler as a porous medium. The model parameters representing the heat transfer coefficient and hydrodynamic resistance of the microchannels are derived from averaging the results of the detailed 3D-CFD simulations of a small region of the microchannels. Upon various modeling optimizations involving the choice of anisotropic porous medium models with a local thermal nonequilibrium and orthotropic heat conduction and hydrodynamic resistance the resulting model was within 10% of measurements.

With the developing architectural ideas involving the use of integrated liquid cooling in 3D ICs, the need for accurate thermal simulation tools has emerged. The 3D-ICE tool was one of such tools for fast simulation of systems with liquid cooling in place [87]. Even though attempts at minimizing the complexity of

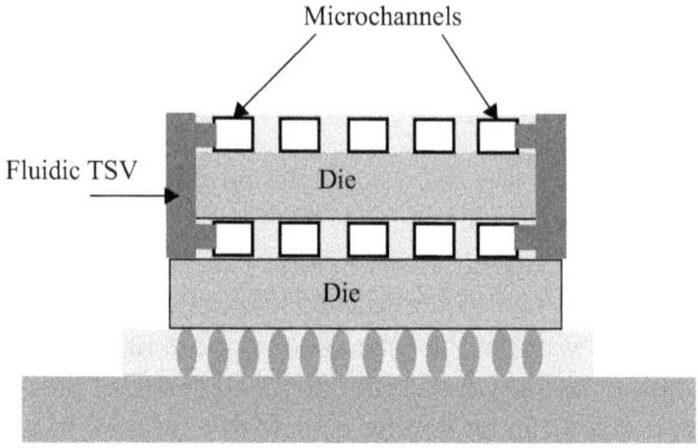

Figure 4.8 Integrated microchannels into a 3D IC using fluidic TSVs. The inlets connecting to the fluidic TSV can be on the topmost layer of the stack or they may also be connected from the bottom through the substrate of the chip

CFD-based methods make this powerful modeling technique more practical, transient thermal simulation often requires orders of magnitude speed-up. 3D-ICE offered such performance benefits (~975-fold speedup compared to a CFD method). An equivalent compact thermal model (similar to those employed for deriving thermal models of IC materials) has been created for a fluid thermal cell and the microchannel. The results of the simulator were found to be within 1.5 K of measurements from a test stack emulating a 3D IC.

Process technologies and fabrication steps also require attention for efficient integration of microchannels in 3D ICs [88]. The outlined process in this study illustrates the major lithography steps, where fabrication of electrical TSVs are followed by microchannels and fluidic TSVs. This study envisions a similar integrated microchannel network for each individual layer as mentioned above and the coolant fluid is supplied to the microchannel heat sinks on the back side of each chip in the stack using fluidic TSVs and pipes connecting the vertical TSV with the horizontal microchannel. Figure 4.8 illustrates a 2-tier stack with one layer of microchannels per die and the fluidic TSVs in place to route the coolant into the respective cooling layer.

Fabrication of fluidic networks within stacked ICs requires innovations taking into account the specific and rather aggressive thermal stress anticipated within heavy duty chips. Specific design guidelines for optimization of fluidic interconnect were introduced to mitigate pressure drop along this avenue [89]. The design problems concern the ability of miniaturized inlets, outlets, and pump structures to maintain the desired flow rate and pressure, particularly at the interfaces that span multiple feature scales. The problems may be thought of as being analogous to designing signal I/O for chips, where drive strength, interfaces, and compatibility also come up as challenges. The patterns of the inlet and outlet ports and interface adapters must be optimized for pressure minimal drop while taking form factor into account.

Creating cost-effective assembly techniques for fluidic networks is crucial for adoption of this cooling technology for high volume manufacturing. Another study focused on this aspect proposing efficient methods for wafer-level integration of electrical and fluidic interconnects into individual chips that can be easily combined later into multi-tier structures [90]. Through wafer fluidic vias and polymer pipes are used to create the microchannels, which are simultaneously assembled with power delivery and signal interconnect. Three alternative fluidic I/O technologies are described including a polymer pipe, a solder-based pipe, and an air-gap C4 I/O. Advantages of solder-based pipes over use of polymers are noted in the relative simplicity of sealing solder-based pipes. They connect with copper rings on the substrate directly without the need of any additional intermediate sealing material. This minimizes the spacing required between layers of the stacks and allows for a more compact implementation. More importantly, unlike polymer-based pipes, solder-based implementations of the fluidic I/O can be performed simultaneously during the same process step as the electrical interconnect simplifying the fabrication cost.

Early works on the modeling and fabrication aspects were followed by studies on the application of emerging liquid cooling technologies on specific computing domains. Naturally, high-performance processor chips have been among the first to be assessed. In one comparison between forced air cooling and liquid cooling applied on 3D multicore processors it was found that liquid cooling yields as additional 45°C reduction of the peak temperature and 12.8% reduction in leakage power [91]. The liquid cooling system was modeled in a similar fashion as the system depicted in Figure 4.8 with integrated channels between layers. Positive influence of liquid cooling on various reliability metrics (electromigration, stress migration, time-dependent dielectric breakdown, negative bias temperature instability, and thermal cycling) were also quantified with architectural simulations combined with a thermal model of the cooling system.

Another study performed thermal analysis of various architectural configurations in a 3D IC package and concluded that some of these architectural enhancements cannot reach their full potential unless aggressive localized active cooling is incorporated. An evaluation of a 3D stacked CPU-DRAM chip with integrated microchannel liquid cooling revealed that the performance of the chip is enhanced by 2.62X thanks to the inclusion of liquid cooling [92]. Liquid cooling essentially enables higher core count, more complex memory controller hierarchy, higher clock frequency for the cores, and increased overall activity subject to the same thermal envelop as a similar chip without liquid cooling.

Run-time management of the cooling system in coordination with other system management tasks for processors is also drawing attention. The dynamic management of liquid cooling via variable flow rate control has been one of such early proposals [79]. A controller has been proposed to adjust the flow rate so that the pump energy can be preserved and the system need not be overdesigned for the worst-case demand. The flow rate control mechanism has been combined with a job scheduler to meet the required cooling demand while minimizing the temperature variation across a 3D IC and the energy consumed by the liquid cooling pump. The required cooling is translated to a target flow rate, which in turn is

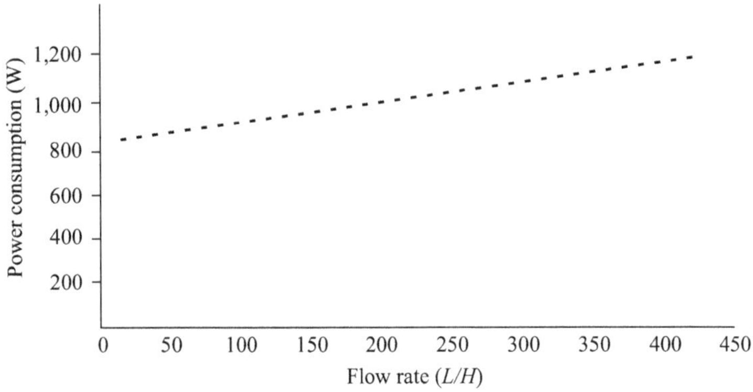

Figure 4.9 The relationship between flow rate and power consumption for a server liquid cooling pump [93]. The data is as advertised by the manufacturer's datasheet and depicts a linear dependency

associated with pump power according to the pump manufacturer's specifications. The power consumption modeled in this particular study is adopted from a pump used for server processor chip cooling and the manufacturer's power specification is depicted in Figure 4.9 [93].

The controller receives feedback from on-chip temperature sensors of the processor cores located in this 3D IC. It is noted in this study that the choice between a predictive versus reactive controller depends on the response time of the pump to updates of its flow rate and the thermal time constant of the underlying IC. In this particular study, the thermal time constant of the 3D IC (~100 ms) is smaller than the latency of changing the flow rate of the pump (~300 ms). Therefore, a predictive mechanism is adopted, where using the temperature sensor feedback of the recent history, decisions on the flow rate is made for the next sample period. The ARMA method has been used to predict the maximum temperature for the next period. For a number of temperature levels the IC is characterized by the minimum flow rate necessary to bring the peak temperature back to a pre-determined threshold, which has been set to 80°C in this study. This information is stored in a LUT and the controller accesses this table to determine the required flow rate at run-time based on the temperature forecast for the subsequent execution period. This study assumes that all microchannels in the system are subject to the same flow rate uniformly. Another approach considers the option of controlling the flow rates of a number of microfluidic channels independently to create a uniform temperature profile over a 3D multicore cache-processor system [94]. The controller relies on system power consumption feedback to estimate future temperature profiles. For this purpose, an autoregressive (AR) prediction with correction is implemented by Kalman filtering to predict and correct the run-time power. Using a compact thermal model, this information is translated to temperature prediction. In response to temperature predictions, fine grain flow rate allocation in the microchannels is performed. On the other hand, a dynamic optimization method to control each individual microchannel would incur significant computational overheads. The microchannels in the cooling system are clustered into groups, each with a dedicated pump, to achieve a trade-off between complexity and optimization efficiency. In this particular study, each control period is about 600 ms and at the start of each period the maximum temperature is predicted for each chip region coinciding with each microchannel cluster. If the estimated temperature is higher than a threshold, the flow rate is increased. If it is less than the threshold, the flow rate is reduced. Each update to the flow rate is proportional to the square of the temperature difference scaled with an empirical factor α as depicted in (4.19).

$$Flow\ Rate[n+1] = Flow\ Rate[n] + \left[1 + \alpha(T_{est} - T_{limit})^2\right] \quad (4.19)$$

The dynamic management of flow rates in mcirochannels can also be combined with design time decisions on their allocation and placement. One approach argues that a non-uniform placement and allocation of microchannels in locations of higher cooling need may result in more effective heat removal than globally uniform distribution of microchannels between all tiers in a 3D IC [95]. Candidate locations for possible placement of microchannels are determined by taking TSV locations through the IC into consideration. This yields a set of possible grid locations for microchannels. The pump power is approximated by the product of pressure drop and flow rate over all N channels in the system, as shown in (4.20).

$$P_{pump} = \sum_{n=1}^{N} flow_n \times \Delta p_n \quad (4.20)$$

Recalling (4.16) defining the relationship between pressure drop, channel geometry, and flow rate, if channel geometry is assumed to be constant over all channels, the flow rate can be expressed in terms of pressure drop. As a result, the total pump power will be a function of pressure drop per channel and the total number of microchannels.

The problem of optimizing the liquid cooling system is then formulated in two ways. First, the problem is treated purely as a channel allocation problem, where fully developed laminar flow, constant channel dimensions, and same constant pressure drop in all channels are assumed. In this case, microchannel allocation (location and number of channels) is an optimization problem subject to maximum allowed temperature within a grid region in the chip, with the objective function being the minimization of total pump power. In the second formulation, pressure drop is also treated as a variable and co-optimized with microchannel allocation.

Mitigation of thermal problems in 3D MPSoCs is also drawing attention as they evolve into a densely integrated collection of a number of high-performance high heat-output chips. Even in the presence of liquid cooling, temperature sensing still plays a crucial role in the overall thermal management paradigms of these systems, which is emphasized by a recent proposal for combined sensor placement and thermal management [96]. In this study, a 4-tier MPSoC, with microchannels etched into the bulk silicon layer of each tier is being modeled. Flow rates of the channels along each tier is managed independently by a dedicated pump at run-time. A state-space representation is utilized to represent the thermal response of the 3D MPSoC, which is decomposed into a grid of cells. The numbers of states in this system corresponds to the grid cells, where temperature is being modeled (estimated). The problem of sensor placement corresponds to determining which of these cell locations (states) will contain a temperature sensor. A model order reduction technique is utilized to eliminate those states with minimal contribution to the system's response using the Gramian-based balancing of state-space realizations [97] by quantifying the role of a grid cell in the heat propagation through this system with an energy metric. The same principle is used to identify the most discerning cells towards thermal events and a minimum number of these locations are identified for temperature sensor placement while maximizing the observability. A target value for the rank of the observability matrix of the system is given and sensor locations are added to the solution in a greedy manner until this target level of observability is reached. At run-time, the reduced order thermal model is used to estimate a thermal map of the system, where temperature sensor feedback is applied on a state estimator (techniques discussed in Section 3.3.1.2 would be applicable) to evaluate the cooling need in various locations of the IC. Flow rate of the liquid cooling system is adjusted to the minimum pump speed that is sufficient to meet this need. This dynamic management scheme also coordinates the voltage and frequency scaling of the chip simultaneously. A closed loop control optimization problem is solved for each discrete control period to determine the voltage, clock frequency, and flow rate of all pumps.

Latest studies combine a multitude of design aspects including thermal-aware floorplanning, TSV integration, high-speed thermal simulation, and liquid cooling [98–101]. These studies involve an increasingly larger variety of system including stacked memory, GPGPUs, photonic circuits, power circuits, etc. and deal with both chip-level and package-level cooling. Liquid cooling technologies will maintain their significant impact at macroscale for datacenters and supercomputers (a brief discussion on cooling at this level of abstraction will be presented in Chapter 5) for cabinet cooling. At the same time, the high level of activity on development of chip-level active cooling solutions, as discussed in this chapter, indicates intriguing new solutions for the future of on-chip active cooling.

4.2.3 Direct liquid cooling

Finally, an alternative that does not require tubes and circulation with a pump is the direct liquid cooling technology [102–105]. Direct liquid cooling systems allow the coolant to come in physical contact with the electronics. Therefore, an electrically isolating material needs to be used as a coolant, hence, the choice is some type of dielectric. A dielectric liquid is injected over the hot surface, the coolant evaporates absorbing heat. Use of a dielectric ensures that the chip and surrounding signal and power lines can come in direct contact with the coolant while maintaining electrical isolation. The perfluorocarbons and hydro-fluoro-ethers are suitable candidates for the coolant liquid [102]. Direct liquid cooling can be implemented by immersion or by jet impingement directed at a specific spot on the hot surface.

4.3 Thermoelectric cooling

By making use of specially designed miniature heat pumps, active cooling technologies can provide site-specific and on-demand cooling, promising new opportunities in cost-effective cooling. More profoundly, the capability of providing tunable cooling at a fine granularity has significant implications on the thermal-aware design of the ICs and systems. Potentially, the active cooling system, the thermal monitoring system, and the architecture-level thermal management mechanisms can operate synergistically to achieve enhanced performance under a safe operating temperature. Examples of methods and technologies to realize this paradigm using liquid cooling have been discussed in the previous section. Another promising on-chip active cooling technology leverages the thermoelectric phenomenon.

The TEC is one of the most accessible technologies with a longstanding history of existing bulk devices in practical use. While devices prior to 1950s were not of high efficiency due to poor pairing of known metals, the birth of a new era for thermoelectric materials is largely attributed to the introduction of compound tellurides (Bi_2Te_3) [106] and alloying techniques in mid-1950s [107].

In more recent times discrete TEC elements have found use in systems, ranging from cooling power electronics and lasers to humans [108–112]. However, integrating such discrete TEC elements into the chip package and further on-chip integration for hotspot cooling is generally not an option, since the heat pumping capability of the discrete devices is limited due to the use of thick bulk thermoelectric materials. Reducing the thickness of the thermoelectric materials help increase the heat pumping capability. For instance, it is shown that if the thickness of the TEC devices can be made to be less than 50 μm, it is possible to achieve the heat pumping capability of 500 W/cm^2 [113]. During the last few years, significant advances in thin-film TEC devices have been reported in the literature. A 5 μm thick Bi_2Te_3 superlattice TEC device has been introduced, which is estimated to achieve 500 W/cm^2 of cooling capability [114]. Monolithic integration of conventional bulk materials on silicon substrate has been investigated, resulting in a device deposited as 20 μm legs using the IC fabrication process [115]. A maximum cooling capability of 100 W/cm^2 was reported. An embedded TEC, which is about 100 μm thick and can be mounted on the heat spreader illustrates one possible package-level integration solution [116]. More recently, the integration of thermoelectric coolers fabricated from nanostructured Bi_2Te_3-based thin-film superlattices into electronic packages has been demonstrated [113]. An on-demand cooling swing of 7.3°C is reported at the targeted region on a silicon chip with a local heat flux as high as 1300 W/cm^2.

These early developments towards package-level and on-chip integration of TEC elements paved the way for a large body of work on design automation for TEC optimization and chip-level integration problems. In the following, we will first review the basic operating principles of TEC elements and then discuss the latest contributions focusing on the on-chip cooling and thermal management aspects of this exciting technology.

4.3.1 Principles of operation and performance metrics for TEC devices

A TEC device is typically composed of a couple of dissimilar semiconductor strips connected electrically in series and thermally in parallel (as depicted in Figure 4.10 (a)). The principle behind the thermoelectric cooler is the Peltier effect: when an electrical current is sent through the strips, heat is absorbed at one side and released at the other side. Denoting the heat flux absorbed from the cold side as Q_c and dissipated from the hot side as Q_h at temperatures T_c and T_h, respectively, the following equations describe the principle of thermoelectric cooling:

$$Q_c = \alpha i T_c - \frac{1}{2} r i^2 - k(T_h - T_c) \tag{4.21}$$

$$Q_h = \alpha i T_h - \frac{1}{2} r i^2 - k(T_h - T_c) \tag{4.22}$$

i is the electrical supply current of the TEC device; α is the Seebeck coefficient of the TEC device; r and k are the electrical resistance and thermal conductance of the device, respectively. The first term in both (4.21) and (4.22) describes the Peltier cooling effect. The second term is due to Joule heating that occurs within the TEC device – half of the Joule heat is dissipated at the cold side and the other half at the hot side. The third term is contributed by heat conductance from the hot side to the cold side. The input power of a TEC device is equal to

$$P_{TEC} = Q_h - Q_h = r i^2 + \alpha i(T_h - T_c) \tag{4.23}$$

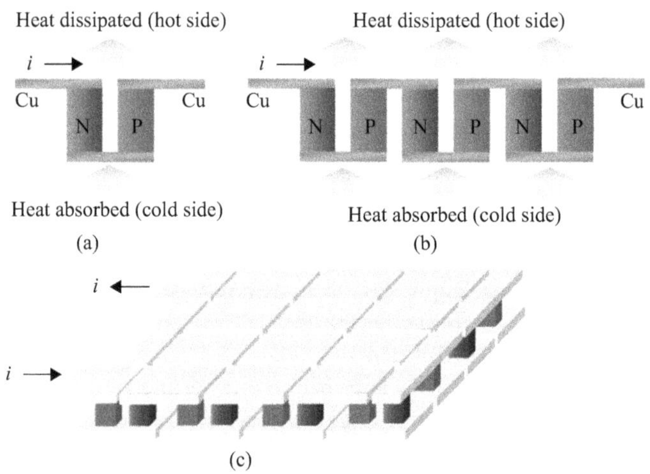

Figure 4.10 (a) The side view of a single TEC device, (b) multiple TEC devices connected electrically in series and thermally in parallel, and (c) the 3D view of a 4 × 4 array of thin-film TEC devices

In steady state, the input power of the TEC devices will be converted to heat in the chip package before being dissipated to the ambient. Hence, an excessive deployment of TEC devices and/or improper setting of the TEC supply current levels could lead to the overheating of the chip package.

Although the above equations are widely used in the literature, they have omitted the Thompson effect, which accompanies the Peltier effect [117, 118]. The Thompson effect is caused by the dependence of the Seebeck coefficient on temperature. However, the Thompson effect can be accounted for by substituting α with $(\alpha_H + \alpha_C)/2$, where α_H and α_C are the Seebeck coefficients at the highest and lowest allowable operating temperature of the system, respectively.

In order to enhance the cooling effect, multiple TEC devices can be connected electrically in series and thermally in parallel as shown in Figure 4.10(b). Figure 4.10(c) illustrates the 3D view of a 4×4 array of superlattice thin-film TEC devices. Thin-film TEC devices occupy small areas. For instance, a 7×7 array of thin-film TEC devices has a lateral dimension of about 3.5 mm $\times$ 3.5 mm [113].

Figure 4.11 illustrates the on-chip thermoelectric cooling system and its integration into the chip package. The cooling system contains a set of TEC devices. These devices are immersed in the TIMs layer, lying between the die and the heat spreader layer. The TEC devices are driven by an external electrical current source.

The performance of a TEC device is characterized by two metrics. The first is the figure of merit of the cooler Z, which is often combined with the absolute temperature T into a dimensionless quantity. The figure of merit of a thermoelectric cooler is defined as:

$$Z = \frac{\alpha^2}{\rho\kappa} \tag{4.24}$$

α is the Seebeck coefficient of the thermoelectric material, and ρ and κ are the electrical resistivity and thermal conductivity of the material, respectively. The temperature range of interest for an active cooling device for ICs is 300–400 K. While higher figure of merit thermoelectric materials under 300 K range are available, the main challenge in recent technological innovations for chip cooling has been in achieving thin-film materials with Z larger than one at room temperature. Enhancements to the figure of merit has been so far achieved by manipulating the lattice structures and engineering superlattices, where most notable success for

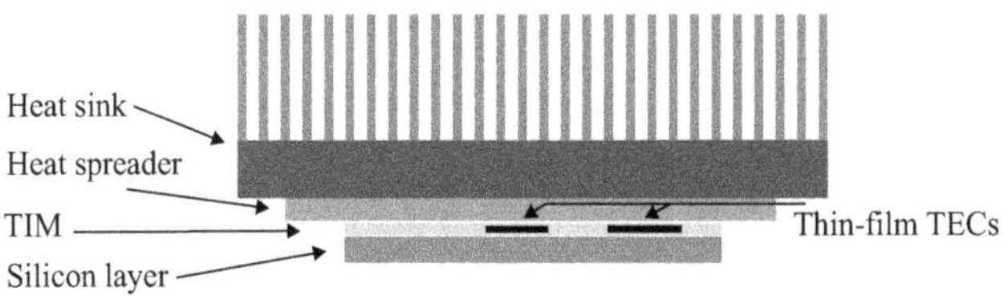

Figure 4.11 Integration of the thin-film TECs into the chip package: the cooling devices can be formed as an array of tiles and embedded into the thermal insulation material between the die and the heat spreader

chip cooling applications have been achieved with the Bi_2Te_3 system [114, 119, 120]. Common methods include creation of quantum dot structures and engineering acoustic mismatch in the superlattice for blocking phonons and transmitting electrons. A general observation that can be made by inspecting (4.21) and (4.24) is that materials with lower higher Seebeck coefficients, lower electrical resistivity, and lower thermal conductivity would yield the desired outcome for both metrics.

The coefficient of performance (COP) of a thermoelectric cooler is defined by the amount of heat pumped by the cooler per unit power dissipated by the cooler:

$$COP = \frac{Q_c}{P_{TEC}} \tag{4.25}$$

4.3.2 Design of latest generation on-chip coolers

The first step towards achieving high COP TEC devices for chip cooling has been investigating optimal design for bulk materials that can be attached to discrete heatsinks [121]. An early study evaluated the combination of air fan and thermoelectric cooler for a microprocessor systems to assess cooling performance over a wide range of varying thermal states [122] . The thermoelectric cooler is attached to a remote heatsink on the board and the processor engages the cooler only if the temperature detected at the heatsink exceeds a threshold. Other early works focused on the design parameters of the TEC device. One aspect of optimization is identified as the geometry of the device. Various methods for determining optimal feature sizes for TEC devices used in discrete components as well as in integrated heatsinks have been proposed, including numerical optimization methods to determine the device geometry [123, 124] and analytical frameworks to determine the optimal height of the TEC devices [125]. These methods generally aim to achieve optimal thermal conductivity and electrical resistivity of the TEC device for a given heat dissipation by the IC subject to cooling. Another important parameter is the bias current supplied to the TEC. There is a tight relationship between the TEC bias current and cooling efficiency [112, 126]. For a given heat pumping requirement or temperature difference to be maintained between the cooled system and the ambient, there is an optimal amount of bias current to be supplied into the TEC. This is a crucial design parameter for optimal TEC operation, both for bulk and thin-film devices. Recent proposals for optimal configuration of thin-film TEC devices, specifically for chip cooling applications will be discussed in more detail in later sections.

The final aspect of TEC design is developing efficient models for evaluation of design choices and TEC-chip-package integration. This entails analytical models for the thermal and electrical response of TEC devices within the environment that they are deployed. Equivalent thermal circuit models, SPICE-like models, and computational models for both steady state and transient analysis of TEC devices have been proposed [127–129]. These earlier models involve complicated circuit elements such as voltage-controlled-voltage-sources and temperature-dependent heat sources. A recent model provided a simpler representation using only thermal conductors and heat sources [130]. In the following, major features of an equivalent TEC model will be illustrated using this model.

4.3.2.1 Thermal model of a chip package including a thermoelectric cooler

An equivalent thermal model for the TEC device can be constructed based on (4.21) and (4.22) as depicted in Figure 4.12. The two nodes in the network represent the hot and the cold sides of the device. The cold side faces the chip being cooled and the hot side faces the heatsink. A thermal conductor k connecting the two nodes models the term $k(T_h - T_c)$. The Joule heating effect can be described by two heat sources connected to the hot side and cold side nodes, each having the magnitude of $ri^2/2$. The Peltier heat $\alpha i T_c$ absorbed at the cold side can be described by a thermal conductor αi connecting the cold side and the ground node, which represents the absolute zero temperature. Likewise, the Peltier heat $\alpha i T_h$ released at the hot side can be modeled using a negative thermal conductor $-\alpha i$ connecting the hot side and the ground. The values of these two conductors are modulated by the supply current, embodying the dependence of effective thermal resistance on bias current. Hence, in Figure 4.12, arrows were used to indicate that they are "tunable conductors." In addition, two thermal conductors g_h and g_c are placed at the hot/cold nodes to account for the contact thermal resistance between the hot/cold side and the rest of the package. It should be noted that this network represents a thermal flow circuit and the physical TEC bias current i does not actually flow through any of these "conductances."

Based on this equivalent model a framework to express TEC operation and its interaction with the environment can be constructed. Figure 4.13(a) illustrates an example for an equivalent network of a package with TEC devices. This network is a connected conductor network containing both fixed and tunable conductors (depicted in Figure 4.13(a)). The fixed conductors must have positive values. In Figure 4.13(a), the conductors colored in bold are the fixed conductors. These fixed

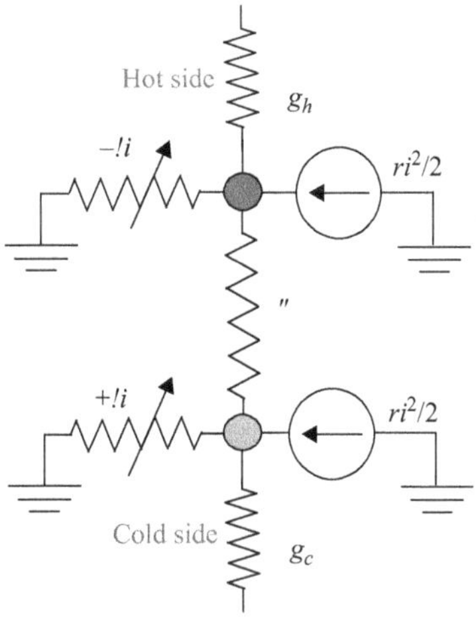

Figure 4.12 Equivalent thermal circuit representing a thermoelectric cooling device

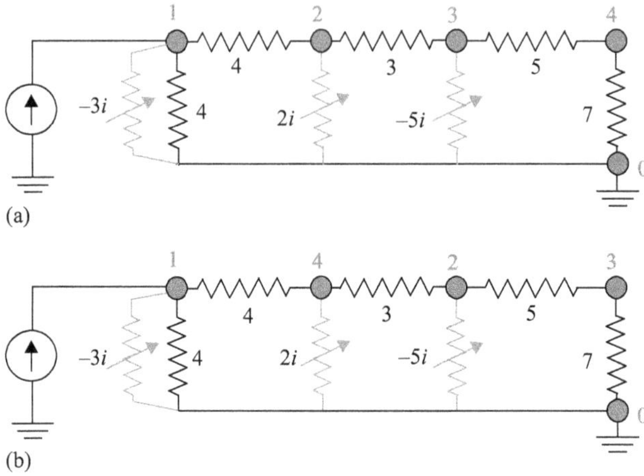

Figure 4.13 A simple example of the generalized network representation of a system with TEC devices deployed: (a) the nodes are labeled such that the corresponding matrix **G** *is a fully irreducible matrix and (b) nodes are labeled such that the corresponding matrix* **G** *is not a fully irreducible matrix*

conductors form a connected sub-network. A special node labeled 0 in this sub-network is defined as the ground node. The tunable conductors can be placed between any node in this sub-network and the ground node. In Figure 4.13(a), they are depicted in gray with arrows. The value of the tunable conductor connected to the kth node is $-\alpha_k i$, where i is a real number representing an external tuning force (in the case of thermoelectric cooling system, i is the supply current). All the tunable conductors share the same value of i if the TEC device is biased with a single current source. The tunable conductors can have either positive or negative values. Further, in order to make the analysis applicable to the thermoelectric cooling system, we require that at least one of tunable conductors have a negative value. That is, at least one α_k has to be positive.

The thermal steady state of the network can now be computed by extending the nodal analysis technique. Let us use notations SIL, HOT, and CLD to denote the set of nodes within the general equivalent thermal network representing a package, where SIL corresponds to all nodes in the silicon layer, HOT is the set of nodes at the hot side of the TEC device, and CLD are the nodes at the cold side of the TEC device. A vector $\boldsymbol{p}$ represents the input heat power at each node, where p_k equals the heating power of the silicon grid points, if $k \in SIL$, $ri^2/2$ if $k \in HOT \cup CLD$, and 0 otherwise. Nodes representing grid locations within the insulation layers, heat spreader, and the heatsink do not generate any power/heat. Then, the steady-state temperature profile θ is related to the power profile $\boldsymbol{p}$ by a system of linear equations

$$(\boldsymbol{G} - i\boldsymbol{D})\theta = \boldsymbol{p}, \tag{4.26}$$

where matrices $\boldsymbol{G}$ and $\boldsymbol{D}$ are defined as follows:

$$\boldsymbol{G}=\begin{pmatrix} \sum\limits_{0\le l\le n} g_{1l} & -g_{12} & \cdots & & -g_{1n} \\ \cdots & & & & \cdots \\ -g_{kl} & \cdots & \sum\limits_{0\le l\le n} g_{kl} & \cdots & -g_{kn} \\ \cdots & & & & \\ -g_{n1} & -g_{n2} & & \cdots & \sum\limits_{0\le l\le n} g_{nl} \end{pmatrix} \quad \boldsymbol{D}=\begin{pmatrix} \alpha_1 & 0 & \cdots & & 0 \\ & \alpha_2 & \cdots & 0 & \\ & & \alpha_k & & \\ & 0 & \cdots & \cdots & \\ 0 & & \cdots & 0 & \alpha_n \end{pmatrix} \tag{4.27}$$

Here g_{kl} is the thermal conductance between node k and node l (g_{kl} is zero if node k and node l are not adjacent in the network); α_k equals to $+\alpha$ if $k \in HOT$, $-\alpha$ if $k \in CLD$, and 0 otherwise, and α accounts for the transfer of Peltier heat in the network.

Matrix $\boldsymbol{G}$ has several important properties that can be leveraged for a formal treatment of the TEC optimization framework. For instance, the labeling in Figure 4.13(a) leads to a fully irreducible matrix $\boldsymbol{G}$, while the labeling in Figure 4.13(b) does not. Without loss of generality, in the following, we assume that the network modeling the chip package is properly labeled such that matrix $\boldsymbol{G}$ is fully irreducible, and the element at the upper-left corner of matrix $\boldsymbol{D}$ is positive.

The network model depicted in Figure 4.13(a) contains only resistive elements and current sources, which would limit it to steady-state analysis. Most current thermoelectric chip cooling systems are being envisioned to have the TEC devices constantly "ON," biased at a particular current level i. In that case, the cooling system need not be designed to respond to transient events, where a steady-state analysis would suffice. On the other hand, future thermoelectric cooling systems may be controlled dynamically in a synergistic fashion as part of the DTM mechanisms. The thermal models for those TEC devices would also include thermal capacitance components in the equivalent circuit network.

4.3.3 A theoretical analysis framework for thermoelectric coolers

The thermal response of the system including TEC devices shown in (4.26) facilitates a comprehensive theoretical framework based on the extension of the classical eigenvalue/eigenvector theory in linear algebra. In this section, the derivation of this framework will be presented. This framework provides important insights to the theoretical performance limits of TEC-based chip cooling [130–132].

The notations in the following discussion are fixed as follows unless stated otherwise explicitly:

1. $\mathbf{M}_k$ is a sub-matrix formed by keeping the first k rows and k columns of the given $n \times n$ matrix M.
2. For a vector $\boldsymbol{v} = (\boldsymbol{v}_i)$, we have
   (a) $\boldsymbol{v} \geq 0$ if and only if $\boldsymbol{v}_i \geq 0$ for all i;
   (b) $\boldsymbol{v} > 0$ iff $\mathbf{v}_i \geq 0$ for all i; and at least one of $\mathbf{v}_j > 0$ for some j;
   (c) $\boldsymbol{v} >> 0$ iff $\boldsymbol{v}_i > 0$ for all i.

3. $\boldsymbol{A}(i) = \boldsymbol{G} - i\boldsymbol{D}$, for some $i \in R$, $\boldsymbol{G}$, and $\boldsymbol{D}$ are $n \times n$ matrices. For simplicity, we use $\boldsymbol{A}$ for $\boldsymbol{A}(i)$. n is the total number of nodes (except for the ground node) in the network.
4. $\boldsymbol{A}_k(i) = \boldsymbol{G}_k - i\boldsymbol{D}_k$, and $i \in R$. For simplicity, we use $\boldsymbol{A}_k$ for $\boldsymbol{A}_k(i)$.
5. $\boldsymbol{H}$ is the inverse of $\boldsymbol{A}$, that is, $\boldsymbol{H}(i) = \boldsymbol{A}(i)^{-1}$ where $\boldsymbol{H}(i) = (h_{kl}(i))$
6. $h_{kl}(i)$ is the entry on the kth row and lth column of $\boldsymbol{H}$.

Definition 1: Let $\boldsymbol{M}$ be a given $n \times n$ irreducible matrix. If $\boldsymbol{M}_k$ is irreducible for each k, then $\boldsymbol{M}$ is called *fully irreducible*.

Definition 2: An $n \times n$ real matrix $\boldsymbol{M}$ is *positive definite* if $\boldsymbol{x}^{\mathrm{T}}\boldsymbol{M}\boldsymbol{x} > 0$ for all non-zero vectors $\boldsymbol{x}$ with real entries.

Definition 3: A *Stieltjes matrix* is a real symmetric matrix with non-positive off-diagonal elements [133].

Lemma 1: Given a generalized conductor network containing at least one negative conductor, it is always possible to label the nodes such that (1) matrix $\boldsymbol{G}$ is fully irreducible, and (2) α_1, the element at the upper-left corner of matrix $\boldsymbol{D}$ is positive.

Proof: Since the chip package contains at least one TEC device, its thermal model contains at least one negative thermal conductor. Let us label the non-ground end node of a negative conductor as the first node, the element at the upper-left corner of matrix $\boldsymbol{D}$ would be positive. Starting from this node, we continue to label the nodes such that the latest labeled node is adjacent to at least one of the nodes that have been labeled. This labeling order ensures that the sub-matrix formed by the first k row and k column of $\boldsymbol{G}$ corresponds to a connected component. Hence, $\boldsymbol{G}$ is fully irreducible.

Theorem 4.1. Matrix $\boldsymbol{G}$ in (4.26) is a fully irreducible positive definite Stieltjes matrix [134].

Using the fundamental definitions presented above, important properties of a TEC device can be analyzed formally. The relationship between cooling efficiency and bias current is one such crucial parameter. In the following, the relationship between heat pumping capability of a TEC device and its bias current will be discussed using this framework.

4.3.3.1 The thermoelectric cooling system thermal runaway phenomenon

There exists an upper limit λ for the supply current of the TEC devices, such that any supply current level larger than λ would cause thermal runaway. Mathematically, as the supply current approaches λ, the temperature of each TEC tile approaches infinity. The theory of inverse-positive matrix [133] can be leveraged to show the existence of the upper limit λ.

Theorem 4.2. Given a positive definite irreducible Stieltjes matrix $\boldsymbol{G}$ and a real diagonal matrix $\boldsymbol{D}$ with at least one positive element, λ is defined as follows:

$$\lambda = \inf\left\{ \frac{\theta^{\mathrm{T}} \boldsymbol{G} \theta}{\theta^{\mathrm{T}} \boldsymbol{D} \theta} \middle| \theta^{\mathrm{T}} \boldsymbol{D} \theta > 0 \right\} \tag{4.28}$$

With $\boldsymbol{A} = \boldsymbol{G} - i\boldsymbol{D}$, we have the following:

For any $i \in [0, \lambda)$, $\boldsymbol{A}$ is positive definite. When $I = \lambda$, $\boldsymbol{A}$ is singular and semi-definite. For any $i \in [\lambda, +\infty)$, $\boldsymbol{A}$ is not positive definite.

Theorem 4.3. Given matrix $\boldsymbol{A}$ and real number λ as defined in Theorem 4.2 and denoting

$\boldsymbol{H} = \boldsymbol{A}^{-1}$, for any $1 \leq k,\ l \leq n$, we have

$$\lim_{i \to \lambda -} h_{kl}(i) = +\infty$$

Figure 4.14 depicts $h_{kl}(i)$ as a function of i. Note that according to (4.26):

$$\theta = \boldsymbol{A}^{-1} \cdot \boldsymbol{p} = \boldsymbol{H} \cdot \boldsymbol{p}$$

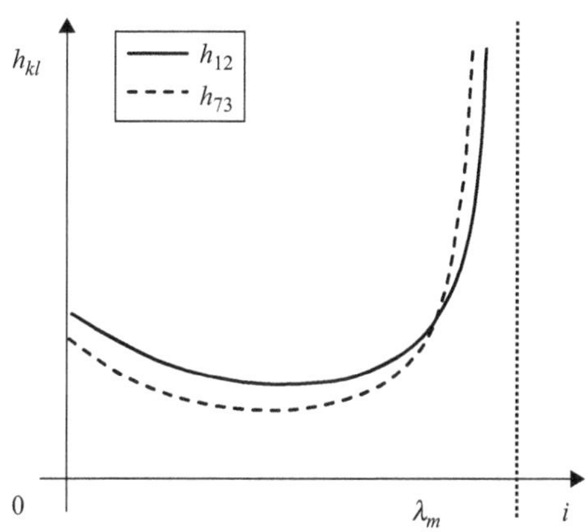

Figure 4.14 Behavior of hkl(i) at different nodes in the system as functions of the TEC supply current i

Thus, the physical interpretation of h_{kl} is the temperature of node k if a unit of power is input at node l. Theorem 4.3 reveals that if there is one node in the network that has non-zero input power, then, as i approaches λ, the temperature of each node approaches infinity, indicating thermal runaway.

The physical interpretation of the thermal runaway is as follows. λ represents the input current level which causes the active cooling system to have zero heat pumping capability since Peltier cooling is offset by the ohmic heating and heat conduction. In the thermoelectric literature, this occurs when the COP of the thermoelectric cooler becomes zero [135]. Similar situations have been investigated but under different boundary conditions [135]. This behavior also reveals that methods for optimization of TEC parameters for chip cooling should restrict the search of the current level that minimizes the maximum silicon layer temperature within $[0, \lambda)$.

4.3.4 Using TEC devices for thermal management of ICs

Designing the overall architecture and configuring the operating parameters of integrated TEC devices for chip cooling involves unique constraints and design trade-offs. First, integrating an excessive number of TEC devices increases the cost of the chip package. Thickness as well as the cross sectional area should be determined with a cost-effective deployment strategy.

Also, supplying the TEC devices with a large electrical current could lead to the overheating of the chip package. As discussed in Section 4.3.3, although the heat pumping capability of the TEC devices increases with their supply current at first, their power consumption increases at a faster pace. Each system (according to its heat dissipation profile, thermal conductance, and TEC materials of choice) possesses a unique optimal point of operation. Beyond that optimal point, the performance of the TEC device degrades exponentially.

Finally, dedicated current supply pins are needed in order to deliver the supply currents from the external sources to the integrated TEC devices. The pin grid array on the modern high-performance microprocessors is quite crowded. This situation would only become more severe in the future, since from generation to generation, the width of the data/address I/O buses increases, while the die size of each processor core keeps decreasing. Thus, the room for extra pins is scarce. Furthermore, within the package, the interconnects bridging the pins and the on-die I/O ports have been very dense. Note that the typical supply current for a TEC device is in the order of a few Amperes. Thus, the interconnect carrying the supply currents have to be made wide enough to suppress the Joule heating effect. However, a wide interconnect would occupy a large routing area. Hence, multiple TEC devices need to be chained together and driven by one single pin to minimize the number of extra pins required.

4.3.4.1 Allocation and optimization for TEC devices in chip packages

An early proposal for a design automation approach to TEC-based chip cooling investigated the problem of identifying the optimal bias current to for a TEC device such that the peak temperature of the target chip is maintained below a safe threshold at all times [131, 132]. Given a $p \times q$ array of tiles representing the die area where each tile has the same area as a TEC device, and the worst-case power consumption of each tile based on profiling of the target chip, the problem and its objective function is formulated as follows:

Determine: (1) The minimal set of tiles that needs to be covered by the TEC devices, and (2) the optimal supply current of the TEC devices;

Objective: The peak steady-state temperature of the silicon layer does not exceed the maximal allowable temperature.

A greedy iterative algorithm is proposed to tackle this problem as follows. With thermal simulation the set of tiles T where the die temperature exceeds the maximum allowable operating temperature θ_{max} is identified first. These tiles are covered with TEC devices. Then, a subroutine is invoked to compute the supply current of the TEC devices that minimizes the peak temperature of the tiles in the silicon layer for the given TEC deployment. Adding more TEC devices into the package has two consequences: the temperatures of the tiles that are covered by these TEC devices may decrease; however, the temperatures of other tiles might increase since the new set of TEC devices dissipate an extra amount of heat in the package.

For each partial deployment of TEC devices to tiles above the die area, a special subroutine computes the supply current of the TEC devices, which minimizes the maximum temperature of the tiles.

This problem can be formulated as follows:

$$\textit{minimize } \textit{max}\{\theta_k(i) \mid \forall k \in \textit{SIL}\}$$

$$\textit{subject to } (\boldsymbol{G} - i\boldsymbol{D})\theta = \boldsymbol{p}, \quad i \geq 0$$

The observation on the theoretical limit λ for the maximum feasible range of current values defines the search space for the optimal supply current. Furthermore, it has been shown that the maximum temperature of the die tiles is a convex function of i over this range $[0, \lambda)$ [130, 132]. Leveraging this outcome, the upper limit is determined and then the gradient descent method is used to perform a search within this range to solve a convex optimization problem minimizing θ_{max}. This framework is then extended to handle a case where sub-regions of the TEC array can be supplied by distinct dedicated current pins. This allows for a larger optimization space to cater to the cooling needs of the underlying chip.

The problem formulation remains similar, except for the replacement of the scalar current variable with a vector $\boldsymbol{i}$ to hold the supply current of each individual pin. Furthermore, the problem has been enhanced with the contribution of TEC self-heating to the total chip power dissipation. Now, the power profile of the overall network depends on the input currents i, since the nodes representing the hot/cold sides of the TEC devices dissipate Joule power. A vector $\boldsymbol{p}(i)$ is used to hold the power dissipation of each node. Then, the steady-state temperature profile θ can be computed by solving the following set of linear equations

$$(\boldsymbol{G} - \boldsymbol{D}(\boldsymbol{i}))\theta = \boldsymbol{p}(i) \tag{4.29}$$

where matrices $\boldsymbol{G}$ and $\boldsymbol{D}(\boldsymbol{i})$ are defined as follows

$$G = \begin{pmatrix} \sum_{0\leq l\leq n} g_{1l} & -g_{12} & & \cdots & & -g_{1n} \\ \cdots & & & & & \cdots \\ -g_{k1} & \cdots & \sum_{0\leq l\leq n} g_{kl} & \cdots & & -g_{kn} \\ \cdots & & & & & \\ -g_{n1} & -g_{n2} & & & \cdots & \sum_{0\leq l\leq n} g_{nl} \end{pmatrix}$$

$$D(i) = \begin{pmatrix} \alpha_1(i) & 0 & \cdots & & 0 \\ & \alpha_2(i) & \cdots & 0 & \\ & & \alpha_k(i) & & \\ & 0 & \cdots & \cdots & \\ 0 & & \cdots & 0 & \alpha_n(i) \end{pmatrix}$$

Matrix $\boldsymbol{G}$ is defined in the same way as in (4.27). Matrix $\boldsymbol{D}(\boldsymbol{i})$ is a diagonal matrix where $\alpha_k(\boldsymbol{i})$ is non-zero only when node $k \in HOT \cup CLD$. When $k \in HOT$, $\alpha_k(\boldsymbol{i})$ equals to the product of the Seebeck coefficient α and the supply current of the corresponding TEC device; when $k \in CLD$, $\alpha_k(\boldsymbol{i})$ would be the negative of the product of the Seebeck coefficient and the supply current.

Now, the problem of optimal cooling system configuration with multiple dedicated current supplies can be defined as follows.

Given a $p \times q$ array of tiles representing the silicon layer of the chip where each tile has the same area as a TEC device, the worst-case power consumption of each tile, and the number of pins n_{pin} to deliver the TEC supply current;

Determine: (1) The set of tiles that needs to be covered by the TEC devices, (2) the mapping of the pins to the TEC devices, that is, the set of TEC devices supplied by each pin, and (3) the supply current level of each pin;

Objective: The maximum steady-state temperature of the silicon die is minimized for the given worst-case power profile.

This problem has been solved in three phases with a relaxation and rounding strategy. Phase 1 is a relaxation step with the transformation of the original problem such that: (1) each tile is covered by a TEC device, and (2) the number of available pins is equal to the number of tiles. The rationale behind the relaxation is that if the conductivity of the TEC devices is larger than that of the TIM (which is the case in practice), a TEC device with zero supply current would become a thermal conductor which conducts heat better than the TIM. As a result, the optimal solution of the relaxed problem places a lower bound on the achievable minimal peak silicon temperature.

The second and third phases "round" the solution of the relaxed problem to the solution of the original problem. The heat dissipation in the silicon layer is highly uneven. Only a small fraction of the tiles would exhibit high temperatures exceeding the safe limit. Hence, in the solution of the relaxed problem, the supply current of most TEC devices would be zero. These TEC devices are removed from the chip package first. Then, the rest of the TEC devices are grouped into n_{pin} clusters. This step determines the deployment of the TEC devices and the mapping between the pins and the TEC devices.

An illustration of histogram of TEC tiles with non-zero supply current demand for a given target chip is shown in Figure 4.15. In this example, there are four naturally formed clusters in the histogram. To explain these clusters, we note that to the first order of approximation, the supply current of a TEC device is determined by the temperature of its underlying silicon tile. The temperature of a tile in the silicon layer, in turn, is approximately determined by the power consumption of itself and its surrounding tiles. Furthermore, we note that tiles within the same functional unit usually have similar power consumption. The spatial correlation among these components indicates that the supply current levels of the TEC devices covering the same functional unit would be close to each other. They form a cluster in the supply current histogram. The k-means clustering technique has been utilized to group supply currents arising from the second phase into n_{pin} clusters. The TEC devices in one cluster would be chained together and supplied by one dedicated current pin.

In the final phase of the rounding process the proper current level for each pin is determined. A specialized search algorithm is used in both the first and the third

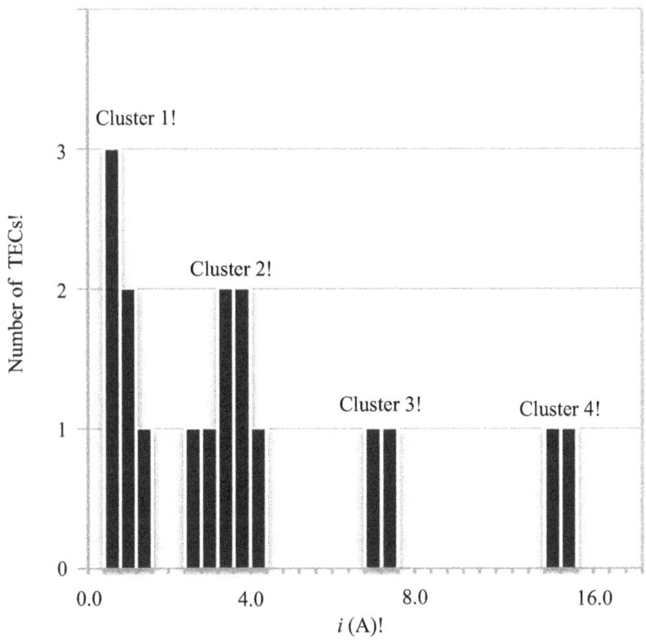

Figure 4.15 The histogram showing the distribution of non-zero TEC currents

phases to determine the current levels for each cluster of TEC devices. The relaxed problem is expressed as follows:

$$\textit{minimize } \theta_{max}$$

$$\textit{subject to } \theta_k \leq \theta_{max} \ \forall k \in SIL$$

$$(\boldsymbol{G} - \boldsymbol{D}(\boldsymbol{i}))\theta = \boldsymbol{p}(\boldsymbol{i})$$

The variable of the above optimization problem is the current vector $\boldsymbol{i}$, which holds the supply current of each pin. θ_k refers to the kth element of vector θ. It represents the temperature of node k. The search space for the current vector for the multi-dimensional case can also be similarly constrained with a convex boundary [132]. This special region $\boldsymbol{R}$ is defined by the space where the matrix $\boldsymbol{G} - \boldsymbol{D}(\boldsymbol{i})$ is positive definite, containing the zero vector $(0, 0, \ldots, 0)^T$.

The starting point of the search is the zero vector $(0, 0, \ldots, 0)^T$. The search algorithm follows the gradient descent method except that before taking each descending step $\Delta\boldsymbol{i}$ from the present point $\boldsymbol{i}$, it needs to be verified that vector $(\boldsymbol{i} + \Delta\boldsymbol{i})$ is within region $\boldsymbol{R}$. If not, the $\Delta\boldsymbol{i}$ step is reduced by half until $(\boldsymbol{i} + \Delta\boldsymbol{i})$ is within $\boldsymbol{R}$. In order to check whether $(\boldsymbol{i} + \Delta\boldsymbol{i})$ is within $\boldsymbol{R}$, it is sufficient to check whether the matrix $\boldsymbol{G} - \boldsymbol{D}(\boldsymbol{i})$ is positive definite. Similar to the single pin case, this can be accomplished using the Cholesky decomposition algorithm [132].

Figure 4.16 presents a case study, where an array of TEC devices are integrated with a microprocessor core. The chip and the TEC layer are divided into a 12×12 grid containing 144 tiles. Figure 4.16(b–d) depict the deployment of TEC devices supplied with a single, two, and three current supply pins, respectively. For the case of four current levels available to the TEC devices (Figure 4.16(d)) the two TEC devices covering floating point adder belong to a cluster driven by Pin4. The two TEC devices cooling the floating point multiplier form another cluster supplied by Pin3. Most of the TEC devices on the integer execution unit and the integer register file form the third cluster driven by Pin2. The rest of TEC devices belong to the fourth cluster supplied by Pin1. The target workload is a floating-point intensive benchmark mix. The floating point functional units have significantly larger power densities compared to the other functional units. Hence, when there are four pins available, the TEC devices covering these two functional units form two individual clusters. When there are only two pins available (Figure 4.16(b)), TEC devices over these two units form one cluster and the rest form the second cluster.

4.3.4.2 Managing TEC integration in computing systems

Following the early results on material, process, and design automation methods for thermoelectric chip cooling as described in the earlier sections, a new and lively domain for research and innovation is forming. New proposals for design and management of integrated TEC devices will be discussed in the following.

A recent experimental study investigated the impact of TEC operating voltage, power dissipation of the heat source (CPU system), and the number of TEC modules, on the overall COP metric of the cooler [136]. This system is at a macroscale, using

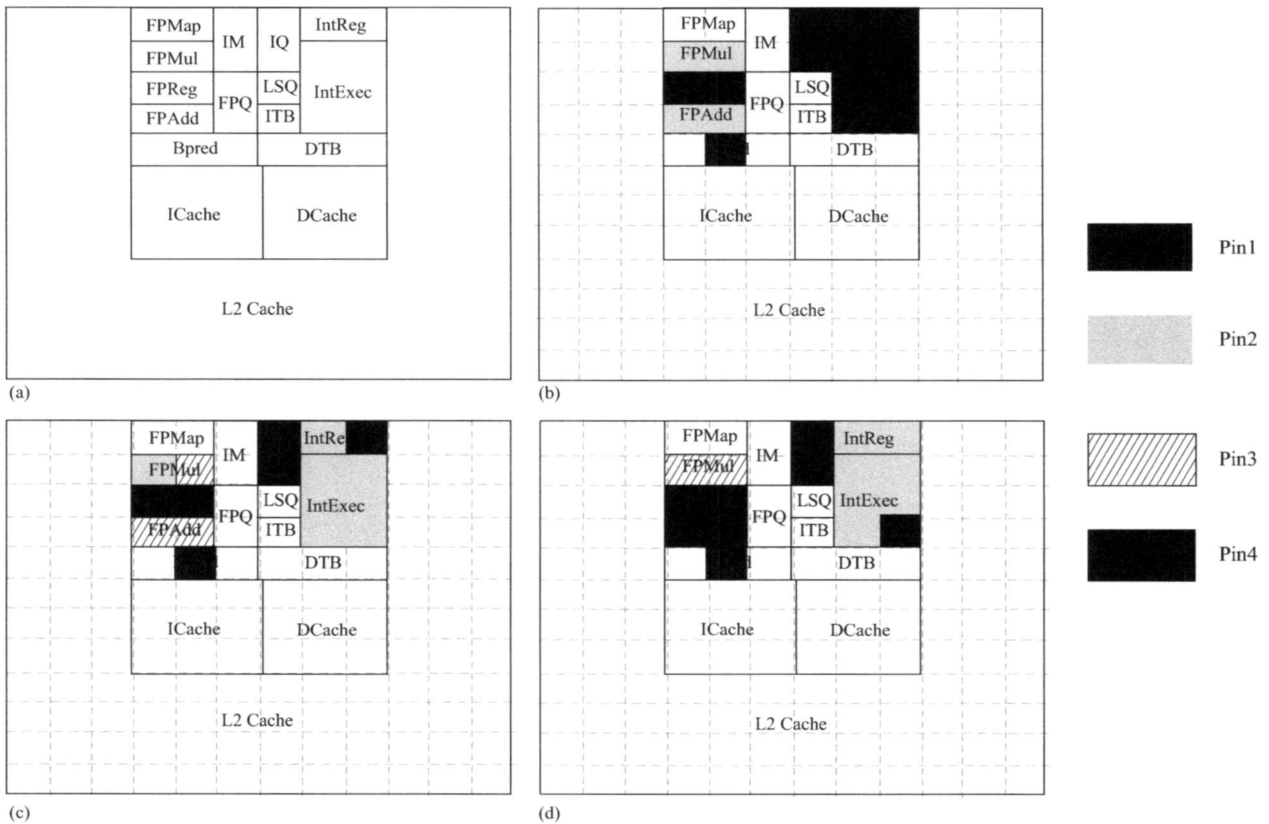

Figure 4.16 *(a) The floorplan of Alpha21364 microprocessor; (b)–(d) the deployment of TEC devices and the pin mapping (2–4 pins) for a floating point workload mix*

TEC devices with dimensions of 40 mm × 40 mm × 3.8 mm attached to the board-level heatsink. The results confirm the similarity of the trends across different scales.

Several other studies evaluate the cooling performance of HY systems comprised of cooling fans and thermoelectric units. One of them evaluated thermoelectric cooling for the emerging cooling needs of notebook computers [137]. This experimental evaluation uses off-the-shelf TEC modules to build a cooling system attached to the air cooled heatsink. A detailed thermal model of the TEC device and finned heatsink was developed. Although a commercially available TEC device was used, the geometry of the heatsink has been optimized as part of this study for optimal air flow and cooling for a typical notebook computer case. The theoretical projections based on the extracted thermal models suggest that this integrated cooler could yield a four-fold to eight-fold improvement in cooling performance relative to passive cooling.

Coordination between two types of cooling systems can be also considered from the viewpoint of total cooling energy spent by the system. Power-aware management of forced-convection and thermoelectric coolers was investigated treating the cooling energy as the primary optimization objective [138]. Providing on-demand cooling with the TEC devices, concentrated only on specific hotspots, relieves significant load off of the primary cooling mechanism. This can lead to a reduced overall cost as it is demonstrated by this study. It can also enable more compact and cost efficient heat-sinks and reduce production cost of passive cooling solutions.

Optimization of TEC specs (i.e., operating current) for a specific platform has been further extended with the notion of performance mode [139]. Particularly, for microprocessor cooling, a distinction is created between cooling for preventive thermal management versus intervention for a thermal emergency. It is found that the setting of the TEC device should be different when aiming for the largest thermal gradient reversal versus when aiming for high COP and energy efficient steady-state cooling.

A large-scale system-level evaluation of thermoelectric cooling in processors on the efficiency of global coolers in a datacenter revealed encouraging conclusions [140]. Datacenter cooling systems are generally designed for ensuring reliability for the processors with highest thermal stress throughout the system, which may often cause overdesign and incurs large energy cost for cooling. Increasing the inlet temperature of the air cooling systems in a datacenter can yield large cost benefits, however, the failure rate due to thermal stress is likely to increase. At the current time, datacenter organizations follow different levels of aggressiveness and some of them rely on the inherent safety margins allocated within processor chips to sustain elevated operating temperatures. Using TEC coolers on select processor chips that are at a significantly larger disadvantage in terms of thermal load (due to location, access to cooling, imbalance in activity, load distribution, etc.) helps to maintain the overall reliability rates while running the datacenter at a warmer air temperature.

4.4 Phase change cooling

PCMs have large heat storage capacities; their temperature gradually increases with heat input up to the point where they undergo a phase transition. At the point of phase transition they exhibit a unique ability to absorb a large amount of heat at a constant temperature as they transition from one phase to another [141–143]. This property is also referred to as the latent heat storage. Solid-liquid PCMs attracted interest in the chip cooling domain and found practical use as an alternative heat-sink material as well as thermal insulation material already. Particularly, in the mobile computing domain, small footprint and light-weight solutions such as the PCM technology could be impactful. They can radically reduce the required volume of conventional heatsink materials and potentially allow a system to run under safe thermal limits without the need of a cooling fan.

Thermal engineers in this domain have hence recognized this potential and several proposals for using PCM-based cooling solutions in hand-held devices have appeared. PCM-based cooling is especially well suited to mitigating bursts of heat output, if such a peak thermal emergency is short-lived. Like all other cooling solutions, several design parameters of PCMs are subject to optimization. Thermal conductivity varies among different PCMs. Hence, the choice and possible enhancements to the materials to improve their conductivity is important for the overall efficiency of the cooling system. For instance a popular organic PCM with a wide range of melting temperatures, paraffin, has low thermal conductivity (0.21–0.24 W/mK). Low thermal conductivity PCMs can be enhanced by introduction of high conductivity stationary structures as inserts [144]. Composites with PCMs are often formed with materials such as copper, aluminum, nickel, and carbon fiber using support structures like fins, honeycomb, wool, etc. Commercially available PCM-based processor heatsinks often use propriety blends that are tuned to melt and absorb heat at a specific, well-controlled temperature range based on the needs of the given platform. Their thermal conductivity is orders of magnitude improved up to several W/mK. PCMs used for skin temperature control of smartphones could be engineered to exhibit latent heat storage in the 36–43°C range to respond to heat-related discomfort by the users [145]. PCM compounds applied to various interfaces between the die, heatspreader, and the heatsink undergo phase change at around 55°C. There are also eutectic metallic PCMs with higher bulk thermal conductivity such as cerrobend (19 W/mK) and gallium (33.7 W/mK) [146]. The main disadvantage of metallic PCMs is their significantly higher weight, which is an important cost issue for electronic packaging such as the peak power input, orientation of package, and melting/freezing times [143]. Modeling is essential for assessing the benefits and limitations of this cooling technology and determining optimal parameters for its design. Various attempts at modeling have been made with varying level of granularity and associated computational complexity from detailed CFD-based models [147] to RC-network equivalent thermal circuits [148]. A more recent model pursues a careful trade-off to enhance prior RC-network-like models to take into account localized heating, non-uniform heat spread and melting point distribution across the PCM surface to achieve an accurate yet computationally efficient model [149]. Using this model, a framework is also constructed for design space exploration of PCM properties. Material thickness and the fraction of metallic reinforcement embedded inside an organic PCM have been considered as two design parameters. Within the feasible range of these parameters, the thermal conductivity varies between 20 W/mK and 106 W/mK. The amount of time that a chip with

PCM-enhanced heatsink spends above a safe thermal limit is observed with simulations of the system. While higher thermal conductivity seems to be the consistently superior solution resulting in the chip remaining hot for shorter periods of time under the same workload, thickness has a non-trivial influence on the thermal response. While higher thickness would be expected to absorb more heat, the temperature of the PCM layer after it melts seems to remain at a higher level for longer. Hence, this analysis cautions for making proper thermal management decisions and optimization of the system considering the expected temperature ranges.

Another study combined CFD modeling and design optimization for PCM-based heat sinks [142]. It establishes that increased power inputs enhance the melting rate and affects the coordination between the PCM layer and the heatsink in terms of their combined thermal conductance until the PCM melts completely. This analysis supports the claim that PCM-based heat sinks could be highly effective for mitigating intermittent thermal peaks.

A cooling system has been proposed for handheld devices and wearable computers with a heat storage unit filled with a PCM [150]. The relative placement and orientation of the heat storage unit with respect to the chips inside the mobile devices has been analyzed to determine highest efficiency operating point for the cooling solution. A related study investigated the design parameters of a PCM-heatsink integration and derived optimal choices for the number of fins on the heatsink, orientation of the heat sources with respect to the heatsink, and the range of power dissipation that could be best managed by the cooling solution [151]. Another use of PCMs in computational sprinting has been demonstrated, where sprinting refers to allowing the system to exceed its TDP rating temporarily to provide instantaneous throughput [152]. In this manner it has been shown that a single core chip in a smartphone system that is thermally constrained for 1 W can be replaced with a 16 core chip, where all 16 cores can be sprinted temporarily to enjoy significant throughput benefits and still operate in the steady state under the original 1 W TDP.

PCM-based materials have already found commercial use. They are likely to be utilized in certain future CPU cooling applications with aggressive thermal stress, where the relatively higher cost (compared to copper heatsinks) is justified. It is also evident from a large number of studies in the mobile computing domain that innovative uses of PCM for skin temperature as well as chip temperature control will continue to emerge.

References

[1] Sauciuc, I., et al., *Thermal performance and key challenges for future CPU cooling technologies.* in *ASME 2005 Pacific Rim Technical Conference and Exhibition on Integration and Packaging of MEMS, NEMS, and Electronic Systems collocated with the ASME 2005 Heat Transfer Summer Conference*. 2005. American Society of Mechanical Engineers.

[2] Torresola, J., et al., Density factor approach to representing impact of die power maps on thermal management. Advanced Packaging, IEEE Transactions on, 2005. **28**(4): pp. 659–664.

[3] Wei, J., Thermal management of fujitsu's high-performance servers. Fujitsu Scientific and Technical Journal, 2007. **43**(1): pp. 122–129.
[4] Vasiliev, L.L., Heat pipes in modern heat exchangers. Applied Thermal Engineering, 2005. **25**(1): pp. 1–19.
[5] *Comair Tutorial: Establishing Cooling Requirements: Air Flow vs Pressure.* http://www.comairrotron.com/establishing-cooling-requirements-air-flow-vs-pressure. 2015, Comair Rotron.
[6] Moss, D., *Guidelines for assessing power and cooling requirements in the data center*. 2005, Dell Power Solutions, August.
[7] Rasmussen, N., Cooling strategies for ultra-high density racks and blade servers. White Paper, 2006. **46**: pp. 1–22.
[8] Patterson, M.K., *The effect of data center temperature on energy efficiency.* in *Thermal and Thermomechanical Phenomena in Electronic Systems, 2008. ITHERM 2008. 11th Intersociety Conference on*. 2008.
[9] Ayoub, R. and T.S. Rosing, *Cool and save: Cooling aware dynamic workload scheduling in multi-socket cpu systems.* in *Design Automation Conference (ASP-DAC), 2010 15th Asia and South Pacific.* 2010. IEEE.
[10] Lefurgy, C., et al., Energy management for commercial servers. Computer, 2003. **36**(12): pp. 39–48.
[11] *Acoustic Noise: Causes, Rating Systems, and Design Guidelines.* http://www.comairrotron.com/acoustic-noise-causes-rating-systems-and-design-guidelines. 2015, Comair Rotron.
[12] Lyon, R.H. and A.E. Bergles, Noise and cooling in electronics packages. Components and Packaging Technologies, IEEE Transactions on, 2006. **29**(3): pp. 535–542.
[13] Shin, D., et al., *Energy-optimal dynamic thermal management for green computing.* in *Proceedings of the 2009 International Conference on Computer-Aided Design*. 2009. ACM.
[14] Wang, Z., et al., *Optimal fan speed control for thermal management of servers.* in *ASME 2009 InterPACK Conference collocated with the ASME 2009 Summer Heat Transfer Conference and the ASME 2009 3rd International Conference on Energy Sustainability*. 2009. American Society of Mechanical Engineers.
[15] Zapater, M., et al., *Leakage and temperature aware server control for improving energy efficiency in data centers.* in *Proceedings of the Conference on Design, Automation and Test in Europe.* 2013. EDA Consortium.
[16] Kim, J., et al., *Global fan speed control considering non-ideal temperature measurements in enterprise servers.* in *Design, Automation and Test in Europe Conference and Exhibition (DATE), 2014*. 2014. IEEE.
[17] Valério, D. and J.S. da Costa, Tuning of fractional PID controllers with Ziegler–Nichols-type rules. Signal Processing, 2006. **86**(10): pp. 2771–2784.
[18] Lee, C. and R. Chen, Optimal self-tuning PID controller based on low power consumption for a server fan cooling system. Sensors, 2015. **15**(5): pp. 11685–11700.

[19] Pradelle, B., et al., Energy-centric dynamic fan control. Computer Science-Research and Development, 2014. **29**(3–4): pp. 177–185.

[20] Han, X. and Y. Joshi, *Energy reduction in server cooling via real time thermal control.* in *Semiconductor Thermal Measurement and Management Symposium (SEMI-THERM), 2012 28th Annual IEEE*. 2012. IEEE.

[21] Yang, B. and X. Ma, *Cooling high-speed chips using intelligent control technology.* in *Automation and Logistics, 2009. ICAL'09. IEEE International Conference on.* 2009. IEEE.

[22] Shin, D., et al., Energy-optimal dynamic thermal management: Computation and cooling power co-optimization. Industrial Informatics, IEEE Transactions on, 2010. **6**(3): pp. 340–351.

[23] Xie, Q., et al., *Adaptive thermal management for portable system batteries by forced convection cooling.* in *Proceedings of the Conference on Design, Automation and Test in Europe.* 2013. EDA Consortium.

[24] Koplow, J.P., *A fundamentally new approach to air-cooled heat exchangers.* Sandia Report No. SANDIA2010-0258, 2010.

[25] Kercher, D.S., et al., Microjet cooling devices for thermal management of electronics. Components and Packaging Technologies, IEEE Transactions on, 2003. **26**(2): pp. 359–366.

[26] Sung, M.K. and I. Mudawar, Single-phase hybrid micro-channel/micro-jet impingement cooling. International Journal of Heat and Mass Transfer, 2008. **51**(17): pp. 4342–4352.

[27] Jiji, L.M. and Z. Dagan, *Experimental investigation of single-phase multijet impingement cooling of an array of microelectronic heat sources.* in *Proceedings of the International Symposium on Cooling Technology for Electronic Equipment.* 1987. Hemisphere Publishing Corporation, Washington, DC.

[28] Wadsworth, D.C. and I. Mudawar, Cooling of a multichip electronic module by means of confined two-dimensional jets of dielectric liquid. Journal of Heat Transfer, 1990. **112**(4): pp. 891–898.

[29] Fabbri, M., S. Jiang, and V.K. Dhir. A comparative study of cooling of high power density electronics using sprays and microjets. Journal of Heat Transfer, 2005. **127**(1): pp. 38–48.

[30] Mahalingam, R., *Modeling of synthetic jet ejectors for electronics cooling.* in *Semiconductor Thermal Measurement and Management Symposium, 2007. SEMI-THERM 2007. 23rd Annual IEEE.* 2007. IEEE.

[31] Mahalingam, R., et al. *Newisys server processor cooling augmentation using synthetic jet ejectors.* in *Thermal and Thermomechanical Phenomena in Electronics Systems, 2006. ITHERM'06. The Tenth Intersociety Conference on*. 2006. IEEE.

[32] Lasance, C.J.M. and R.M. Aarts, *Synthetic jet cooling part I: Overview of heat transfer and acoustics.* in *Semiconductor Thermal Measurement and Management Symposium, 2008. Semi-Therm 2008. Twenty-fourth Annual IEEE*. 2008. IEEE.

[33] James, R.D., J.W. Jacobs, and A. Glezer, A round turbulent jet produced by an oscillating diaphragm. *Physics of Fluids*, 1996. **8**(9): pp. 2484–2495.

[34] Chaudhari, M.B., B. Puranik, and A. Agrawal, Heat transfer characteristics of a heat sink in presence of a synthetic jet. *Components, Packaging and Manufacturing Technology, IEEE Transactions on*, 2012. **2**(3): pp. 457–463.

[35] Sufian, S.F., M.Z. Abdullah, and J.J. Mohamed, Effect of synchronized piezoelectric fans on microelectronic cooling performance. International Communications in Heat and Mass Transfer, 2013. **43**: pp. 81–89.

[36] Ma, H.K., et al., *Thermal performance of multiple piezoelectric-magnetic fans.* in *Semiconductor Thermal Measurement and Management Symposium (SEMI-THERM), 2014 30th Annual. 2014.* IEEE.

[37] Ma, H.K., et al., *Investigation of a piezoelectric fan embedded in a heat sink.* International Communications in Heat and Mass Transfer, 2012. **39**(5): pp. 603–609.

[38] Li, H.-Y., et al., Thermal performance of plate-fin heat sinks with piezoelectric cooling fan. International Journal of Heat and Mass Transfer, 2013. **57**(2): pp. 722–732.

[39] Wait, S.M., et al., Piezoelectric fans using higher flexural modes for electronics cooling applications. Components and Packaging Technologies, IEEE Transactions on, 2007. **30**(1): pp. 119–128.

[40] Acikalin, T., et al., Experimental investigation of the thermal performance of piezoelectric fans. Heat Transfer Engineering, 2004. **25**(1): pp. 4–14.

[41] Jalilvand, A., et al., *Thinner thermal solution module by combination of thin heat pipe and piezo fan.* in *ASME 2011 Pacific Rim Technical Conference and Exhibition on Packaging and Integration of Electronic and Photonic Systems*. 2011. American Society of Mechanical Engineers.

[42] Schacht, R., et al., *Frictionless air flow blade fan for thermal management of electronics.* in *Thermal and Thermomechanical Phenomena in Electronic Systems (ITherm), 2012 13th IEEE Intersociety Conference on.* 2012. IEEE.

[43] Schacht, R., A. Hausdorf, and B. Wunderle, *Feasibility and design study of a frictionless air mover for thermal management of electronics.* in *Journal of Physics: Conference Series*. 2014. IOP Publishing.

[44] Li, A.-T., H.-K. Tseng, and R.-Q. Hsu, Study of an electromagnetic control swing blade blower. Sensors and Actuators A: Physical, 2015. **225**: pp. 61–70.

[45] Peterson, M.S., et al., *Experimental characterization of low voltage field emission from carbon-based cathodes in atmospheric air.* in *ASME 2003 International Mechanical Engineering Congress and Exposition. 2003.* American Society of Mechanical Engineers.

[46] Jewell-Larsen, N.E., *Optimization and miniaturization of electrostatic air pumps for thermal management*. 2004. Master of Science University of Washington, Seattle, WA, USA 2004.

[47] Jewell-Larsen, N.E., et al., *Electrohydrodynamic (EHD) cooled laptop.* in *Semiconductor Thermal Measurement and Management Symposium, 2009. SEMI-THERM 2009. 25th Annual IEEE.* 2009. IEEE.

[48] Ongkodjojo, A., A.R. Abramson, and N.C. Tien, *Design, modeling, and optimization for highly efficient ionic wind-based cooling microfabricated devices.* in *ASME 2010 International Mechanical Engineering Congress and Exposition*. 2010. American Society of Mechanical Engineers.

[49] Hardwidge, B., *Building Extreme PCs: The Complete Guide to Computer Modding*. 2006. O'Reilly Media, Inc., Newton, MA, USA.

[50] Ellsworth, S., *An Expansion on Applied Computer Cooling*, 2012. CalPoly, San Luis Obispo, CA, USA.

[51] Iyengar, M., et al., *Server liquid cooling with chiller-less data center design to enable significant energy savings.* in *Semiconductor Thermal Measurement and Management Symposium (SEMI-THERM), 2012 28th Annual IEEE*. 2012. IEEE.

[52] Huang, W., et al., *Tapo: Thermal-aware power optimization techniques for servers and data centers.* in *Green Computing Conference and Workshops (IGCC), 2011 International*. 2011. IEEE.

[53] Iyengar, M., et al., *Server liquid cooling with chiller-less data center design to enable significant energy savings.* in *Semiconductor Thermal Measurement and Management Symposium (SEMI-THERM), 2012 28th Annual IEEE*. 2012.

[54] Schultz, M., et al., *Experimental investigation of direct attach microprocessors in a Liquid-Cooled chiller-less Data Center.* in *Thermal and Thermomechanical Phenomena in Electronic Systems (ITherm), 2014 IEEE Intersociety Conference on.* 2014. IEEE.

[55] Tuckerman, D.B. and R.F.W. Pease, *High-performance heat sinking for VLSI.* Electron Device Letters, IEEE, 1981. **2**(5): pp. 126–129.

[56] Lee, Y.J., P.S. Lee, and S.K. Chou, Hotspot mitigating with obliquely finned microchannel heat sink—An experimental study. Components, Packaging and Manufacturing Technology, IEEE Transactions on, 2013. **3**(8): pp. 1332–1341.

[57] Zhang, H.Y., et al., Development of liquid cooling techniques for flip chip ball grid array packages with high heat flux dissipations. Components and Packaging Technologies, IEEE Transactions on, 2005. **28**(1): pp. 127–135.

[58] Li, T., et al., A powerful way of cooling computer chip using liquid metal with low melting point as the cooling fluid. Forschung im Ingenieurwesen, 2005. **70**(4): pp. 243–251.

[59] Ma, K.-Q. and J. Liu, Nano liquid-metal fluid as ultimate coolant. Physics Letters A, 2007. **361**(3): pp. 252–256.

[60] Green, C., A.G. Fedorov, and Y.K. Joshi, Fluid-to-fluid spot-to-spreader (F2/S2) hybrid heat sink for integrated chip-level and hot spot-level thermal management. Journal of Electronic Packaging, 2009. **131**(2): p. 025002.

[61] Li, Y., D. Guo, and S.-C. Yao, Thermal-aware microchannel cooling of multicore processors: A three-stage design approach. Journal of Electronic Packaging, 2014. **136**(2): p. 021002.

[62] Phillips, R.J., Forced-Convection, Liquid-Cooled, Microchannel Heat Sinks. 1988, DTIC Document.

[63] Upadhye, H.R. and S.G. Kandlikar, *Optimization of microchannel geometry for direct chip cooling using single phase heat transfer.* in *ASME 2004 2nd International Conference on Microchannels and Minichannels.* 2004. American Society of Mechanical Engineers.

[64] Qu, W. and I. Mudawar, Experimental and numerical study of pressure drop and heat transfer in a single-phase micro-channel heat sink. International Journal of Heat and Mass Transfer, 2002. **45**(12): pp. 2549–2565.

[65] Ryu, J.H., D.H. Choi, and S.J. Kim, Numerical optimization of the thermal performance of a microchannel heat sink. International Journal of Heat and Mass Transfer, 2002. **45**(13): pp. 2823–2827.

[66] Carr, J.D., *Liquid cooling theory and application in systems design.* 2014. DSI Ventures, Tyler, TX, USA.

[67] Kandlikar, S., et al., *Heat transfer and fluid flow in minichannels and microchannels.* 2005. Elsevier US, Philadelphia, PA, USA.

[68] Steinke, M.E. and S.G. Kandlikar, *Single-phase liquid friction factors in microchannels.* in *ASME 3rd International Conference on Microchannels and Minichannels.* 2005. American Society of Mechanical Engineers.

[69] Ellsworth Jr, M.J. and L. Campbell, *Technical Brief: Design Considerations for High Performance Processor Liquid Cooled Cold Plates.* electronics cooling, issue December 22nd, 2011.

[70] Husain, A. and K.-Y. Kim, Shape optimization of micro-channel heat sink for micro-electronic cooling. Components and Packaging Technologies, IEEE Transactions on, 2008. **31**(2): pp. 322–330.

[71] Yong, H., L. Yong Jiun, and Z. Xiaowu, Trapezoidal microchannel heat sink with pressure-driven and electro-osmotic flows for microelectronic cooling. Components, Packaging and Manufacturing Technology, IEEE Transactions on, 2013. **3**(11): pp. 1851–1858.

[72] Sabry, M.M., et al., Greencool: An energy-efficient liquid cooling design technique for 3-d mpsocs via channel width modulation. Computer-Aided Design of Integrated Circuits and Systems, IEEE Transactions on, 2013. **32** (4): pp. 524–537.

[73] Chung-Hao, K., et al., *Optimization of the micro channel ladder shape heat sink.* in *Nano/Micro Engineered and Molecular Systems (NEMS), 2013 8th IEEE International Conference on.* 2013.

[74] Wei, X. and Y. Joshi, Stacked microchannel heat sinks for liquid cooling of microelectronic components. Journal of Electronic Packaging, 2004. **126**(1): pp. 60–66.

[75] Wei, X. and Y. Joshi, Optimization study of stacked micro-channel heat sinks for micro-electronic cooling. Components and Packaging Technologies, IEEE Transactions on, 2003. **26**(1): pp. 55–61.

[76] Ansari, D., A. Husain, and K. Kwang-Yong, Multiobjective optimization of a grooved micro-channel heat sink. Components and Packaging Technologies, IEEE Transactions on, 2010. **33**(4): pp. 767–776.

[77] Li, R.-R., et al., Convective heat-transfer characteristics of a channel with one surface having mini-grooves in the flow direction and a plain surface located at a mini-distance. Components, Packaging and Manufacturing Technology, IEEE Transactions on, 2015. **5**(1): pp. 65–74.

[78] Kandlikar, S.G. and W.J. Grande, Evaluation of single phase flow in microchannels for high heat flux chip cooling—Thermohydraulic performance enhancement and fabrication technology. Heat Transfer Engineering, 2004. **25**(8): pp. 5–16.

[79] Coskun, A.K., et al., *Energy-efficient variable-flow liquid cooling in 3D stacked architectures.* in *Design, Automation & Test in Europe Conference & Exhibition (DATE), 2010.* 2010.

[80] Singhal, V., S.V. Garimella, and A. Raman, Microscale pumping technologies for microchannel cooling systems. Applied Mechanics Reviews, 2004. **57**(3): pp. 191–221.

[81] Iverson, B.D. and S.V. Garimella, Recent advances in microscale pumping technologies: A review and evaluation. Microfluidics and Nanofluidics, 2008. **5**(2): pp. 145–174.

[82] Gillot, C., C. Schaeffer, and A. Bricard. *Integrated micro heat sink for power multichip module.* in *Industry Applications Conference, 1998. 33rd IAS Annual Meeting. The 1998 IEEE.* 1998. IEEE.

[83] Chen, X.Y., et al., *Direct liquid cooling of a stacked MCM.* in *Thermal and Thermomechanical Phenomena in Electronic Systems, 2004. ITHERM'04. The 9th Intersociety Conference on.* 2004. IEEE.

[84] Furmanczyk, M., P. Wilkerson, and A. Przekwas, *Multiphysics modeling of integrated microfluidic-thermoelectric cooling for stacked 3D ICs.* in *Semiconductor Thermal Measurement and Management Symposium, 2003. Ninteenth Annual IEEE.* 2003. IEEE.

[85] Koo, J.-M., et al., Integrated microchannel cooling for three-dimensional electronic circuit architectures. Journal of Heat Transfer, 2005. **127**(1): pp. 49–58.

[86] Alfieri, F., et al., 3D integrated water cooling of a composite multilayer stack of chips. Journal of Heat Transfer, 2010. **132**(12): p. 121402.

[87] Sridhar, A., et al., *3D-ICE: Fast compact transient thermal modeling for 3D ICs with inter-tier liquid cooling.* in *Proceedings of the International Conference on Computer-Aided Design.* 2010. IEEE Press.

[88] Sekar, D., et al., *A 3D-IC technology with integrated microchannel cooling.* in *Interconnect Technology Conference, 2008. IITC 2008. International.* 2008. IEEE.

[89] Tang, G.Y., et al., Integrated liquid cooling systems for 3-D stacked TSV modules. Components and Packaging Technologies, IEEE Transactions on, 2010. **33**(1): pp. 184–195.

[90] King Jr, C.R., et al., *Electrical and fluidic C4 interconnections for inter-layer liquid cooling of 3D ICs.* in *Electronic Components and Technology Conference (ECTC), 2010 Proceedings 60th.* 2010. IEEE.

[91] Jang, H.B., et al., *The impact of liquid cooling on 3D multi-core processors*. in *Computer Design, 2009. ICCD 2009. IEEE International Conference on*. 2009. IEEE.

[92] Serafy, C., A. Srivastava, and D. Yeung, *Unlocking the true potential of 3D CPUs with micro-fluidic cooling*. in *Proceedings of the 2014 International Symposium on Low Power Electronics and Design. 2014*. ACM.

[93] *Datasheet: 12 Volt DC pumps*. Laing GmbH Systeme fur Warmetechnik.

[94] Qian, H., et al., Cyber-physical thermal management of 3D multi-core cache-processor system with microfluidic cooling. Journal of Low Power Electronics, 2011. **7**(1): pp. 110–121.

[95] Shi, B., A. Srivastava, and P. Wang, *Non-uniform micro-channel design for stacked 3D-ICs*. in *Proceedings of the 48th Design Automation Conference*. 2011. ACM.

[96] Zanini, F., D. Atienza, and G. De Micheli, A combined sensor placement and convex optimization approach for thermal management in 3D-MPSoC with liquid cooling. INTEGRATION, the VLSI Journal, 2013. **46**(1): pp. 33–43.

[97] Laub, A.J., et al., Computation of system balancing transformations and other applications of simultaneous diagonalization algorithms. Automatic Control, IEEE Transactions on, 1987. **32**(2): pp. 115–122.

[98] Cuesta, D., et al., Thermal-aware floorplanner for 3D IC, including TSVs, liquid microchannels and thermal domains optimization. *Applied Soft Computing*, 2015. **34**: pp. 164–177.

[99] Liu, X.-X., et al., Parallel thermal analysis of 3-D integrated circuits with liquid cooling on CPU-GPU platforms. Very Large Scale Integration (VLSI) Systems, IEEE Transactions on, 2015. **23**(3): pp. 575–579.

[100] Wan, Z., et al., Co-design of multicore architectures and microfluidic cooling for 3D stacked ICs. Microelectronics Journal, 2014. **45**(12): pp. 1814–1821.

[101] Han, X. and Y.K. Joshi, Compact model-based microfluidic controller for energy efficient thermal management using single tier and three-dimensional stacked pin-fin enhanced microgap. Journal of Electronic Packaging, 2015. **137**(1): p. 011008.

[102] Bar-Cohen, A., M. Arik, and M. Ohadi, Dire6ct liquid cooling of high flux micro and nano electronic components. Proceedings of the IEEE, 2006. **94** (8): pp. 1549–1570.

[103] Estes, K.A. and I. Mudawar, Comparison of two-phase electronic cooling using free jets and sprays. Journal of Electronic Packaging, 1995. **117**(4): pp. 323–332.

[104] Hopton, P. and J. Summers, *Enclosed liquid natural convection as a means of transferring heat from microelectronics to cold plates*. in *Semiconductor Thermal Measurement and Management Symposium (SEMI-THERM), 2013 29th Annual IEEE*. 2013.

[105] Lenke, R.U., M. Christoph, and R.W. De Doncker, *Experimental characterization of immersion-cooled devices at elevated ambient temperatures*. in *Power Electronics Specialists Conference, 2008. PESC 2008*. IEEE. 2008. IEEE.
[106] Goldsmid, H.J. and R.W. Douglas, The use of semiconductors in thermoelectric refrigeration. British Journal of Applied Physics, 1954. 5: p. 386.
[107] Ioffe, A.F., *Semiconductor thermoelements and thermoelectric cooling*. 1957, Infosearch, London, UK.
[108] Huang, B.J., C.J. Chin, and C.L. Duang, A design method of thermoelectric cooler. International Journal of Refrigeration, 2000. **23**(3): pp. 208–218.
[109] Chein, R. and G. Huang, Thermoelectric cooler application in electronic cooling. Applied Thermal Engineering, 2004. **24**(14): pp. 2207–2217.
[110] Berger, P.R., et al., Monolithically Peltier–cooled vertical–cavity surface–emitting lasers. Applied Physics Letters, 1991. **59**(1): pp. 117–119.
[111] Upton, J.F. and S.L. Stein, *Responder Technology Alert Monthly (Oct–Nov 2014)*. 2015, Pacific Northwest National Laboratory (PNNL), Richland, WA, USA.
[112] Min, G., D.M. Rowe, and F. Volklein, Integrated thin film thermoelectric cooler. Electronics Letters, 1998. **34**(2): pp. 222–223.
[113] Chowdhury, I., et al., On-chip cooling by superlattice-based thin-film thermoelectrics. Nature Nanotechnology, 2009. **4**(4): pp. 235–238.
[114] Venkatasubramanian, R., et al., Thin-film thermoelectric devices with high room-temperature figures of merit. Nature, 2001. **413**(6856): pp. 597–602.
[115] Böttner, H., *Thermoelectric micro devices: Current state, recent developments and future aspects for technological progress and applications*. in *Thermoelectrics, 2002. Proceedings ICT'02. 21st International Conference on*. 2002. IEEE.
[116] Snyder, G.J., et al., *Hot spot cooling using embedded thermoelectric coolers*. in *Semiconductor Thermal Measurement and Management Symposium, 2006 IEEE 22nd Annual IEEE*. 2006. IEEE.
[117] Goldsmid, H.J., *Applications of thermoelectricity*. 1960. Methuen.
[118] Chen, W.-H., C.-Y. Liao, and C.-I. Hung, A numerical study on the performance of miniature thermoelectric cooler affected by Thomson effect. Applied Energy, 2012. **89**(1): pp. 464–473.
[119] Harman, T.C., et al., Quantum dot superlattice thermoelectric materials and devices. Science, 2002. **297**(5590): pp. 2229–2232.
[120] Kanatzidis, M.G., Nanostructured thermoelectrics: The new paradigm?†. Chemistry of Materials, 2009. **22**(3): pp. 648–659.
[121] Bierschenk, J. and D. Johnson, *Extending the limits of air cooling with thermoelectrically enhanced heat sinks*. in *Thermal and Thermomechanical Phenomena in Electronic Systems, 2004. ITHERM '04. The 9th Intersociety Conference on*. 2004.
[122] Lundquist, C. and V.P. Carey, *Microprocessor-based adaptive thermal control for an air-cooled computer CPU module*. in *Semiconductor*

Thermal Measurement and Management, 2001. 17th Annual IEEE Symposium. 2001.
[123] Abramzon, B., Numerical optimization of the thermoelectric cooling devices. Journal of Electronic Packaging, 2007. **129**(3): pp. 339–347.
[124] Huang, Y.-X., et al., Geometry optimization of thermoelectric coolers using simplified conjugate-gradient method. Energy, 2013. **59**: pp. 689–697.
[125] Hou, P.Y., R. Baskaran, and K.F. Böhringer, Optimization of microscale thermoelectric cooling (TEC) element dimensions for hotspot cooling applications. Journal of Electronic Materials, 2009. **38**(7): pp. 950–953.
[126] Wang, C.H., J.Y. Zhang, and J. Huang, *Thermal performances analysis of microelectronic chip cooling system with thermoelectric components*. in *Advanced Materials Research*. 2011. Trans Tech Publ.
[127] Gupta, M.P., et al., Ultrathin thermoelectric devices for on-chip peltier cooling. Components, Packaging and Manufacturing Technology, IEEE Transactions on, 2011. **1**(9): pp. 1395–1405.
[128] Sabounchi, P. and A. Heydari, *Lumped capacity thermal circuit diagram of active chip cooling schemes*. in *Semiconductor Thermal Measurement and Management Symposium, 2003. 19th Annual IEEE*. 2003. IEEE.
[129] Mitrani, D., et al., *Lumped and distributed parameter SPICE models of TE devices considering temperature dependent material properties*. in *Thermal Investigation of ICs and Systems, 2007. THERMINIC 2007. 13th International Workshop on. 2007*. IEEE.
[130] Long, J.Y., et al., Theory and analysis for optimization of on-chip thermoelectric cooling systems. IEEE Transactions on Computer-Aided Design of Integrated Circuits and Systems, 2013. **32**(10): pp. 1628–1632.
[131] Long, J., S.O. Memik, and M. Grayson. *Optimization of an on-chip active cooling system based on thin-film thermoelectric coolers*. in *Proceedings of the Conference on Design, Automation and Test in Europe*. 2010. European Design and Automation Association.
[132] Long, J. and S.O. Memik, *A framework for optimizing thermoelectric active cooling systems*. in *Design Automation Conference (DAC), 2010 47th ACM/IEEE*. 2010. IEEE.
[133] Varga, R.S., *Matrix iterative analysis*. **Vol. 27**. 2009. Springer Science & Business Media, New York, NY, USA.
[134] Long, J., *Design of thermal monitoring infrastructures and thermal optimizations for high performance chips*. 2010. Northwestern University, Evanston, IL, USA.
[135] Rowe, D.M., *CRC handbook of thermoelectrics*. 1995. CRC Press, Boca Raton, FL, USA.
[136] Liu, D., et al., Thermoelectric mini cooler coupled with micro thermosiphon for CPU cooling system. Energy, 2015. **83**: pp. 29–36.
[137] Yazawa, K., G.L. Solbrekken, and A. Bar-Cohen, Thermoelectric-powered convective cooling of microprocessors. Advanced Packaging, IEEE Transactions on, 2005. **28**(2): pp. 231–239.

[138] Dousti, M.J. and M. Pedram, *Power-aware deployment and control of forced-convection and thermoelectric coolers*. in *Proceedings of the 51st Annual Design Automation Conference*. 2014. ACM.
[139] Dousti, M.J. and M. Pedram, *Platform-dependent, leakage-aware control of the driving current of embedded thermoelectric coolers*. in *Low Power Electronics and Design (ISLPED), 2013 IEEE International Symposium on*. 2013. IEEE.
[140] Biswas, S., et al., *Fighting fire with fire: Modeling the datacenter-scale effects of targeted superlattice thermal management*. in *Computer Architecture (ISCA), 2011 38th Annual International Symposium on*. 2011.
[141] Ahmed, T., *Fanless Cooling Using Thermal Energy Storage Materials*. 2015, Dalhousie University, Nova Scotia, Canada.
[142] Kandasamy, R., X.-Q. Wang, and A.S. Mujumdar, Transient cooling of electronics using phase change material (PCM)-based heat sinks. Applied Thermal Engineering, 2008. **28**(8): pp. 1047–1057.
[143] Kandasamy, R., X.-Q. Wang, and A.S. Mujumdar, Application of phase change materials in thermal management of electronics. Applied Thermal Engineering, 2007. **27**(17): pp. 2822–2832.
[144] Fan, L. and J.M. Khodadadi, Thermal conductivity enhancement of phase change materials for thermal energy storage: A review. Renewable and Sustainable Energy Reviews, 2011. **15**(1): pp. 24–46.
[145] Sutherland, J., *Formable Phase Change Materials as Latent Heat Sinks for Portable Electronic Devices* 2013, Outlast Technologies LLC: Electronics Protection Magazine.
[146] Sharma, A., et al., Review on thermal energy storage with phase change materials and applications. Renewable and Sustainable Energy Reviews, 2009. **13**(2): pp. 318–345.
[147] Tan, F.L. and S.C. Fok, *Thermal management of mobile phone using phase change material*. in *Electronics Packaging Technology Conference, 2007. EPTC 2007. 9th*. 2007. IEEE.
[148] Tilli, A., et al., *Don't burn your mobile!: Safe computational re-sprinting via model predictive control*. in *Proceedings of the 8th IEEE/ACM/IFIP International Conference on Hardware/Software Codesign and System Synthesis*. 2012. ACM.
[149] Kaplan, F., et al., *Modeling and analysis of phase change materials for efficient thermal management*. in *Computer Design (ICCD), 2014 32nd IEEE International Conference on*. 2014. IEEE.
[150] Tan, F.L. and C.P. Tso, Cooling of mobile electronic devices using phase change materials. Applied thermal engineering, 2004. **24**(2): pp. 159–169.
[151] Fok, S.C., W. Shen, and F.L. Tan, Cooling of portable hand-held electronic devices using phase change materials in finned heat sinks. International Journal of Thermal Sciences, 2010. **49**(1): pp. 109–117.
[152] Raghavan, A., et al., *Computational sprinting*. in *High Performance Computer Architecture (HPCA), 2012 IEEE 18th International Symposium on*. 2012.

Chapter 5
Mitigating thermal events at the system level and above

Heat in computing systems originates from power consumption of active devices and Joule heating of interconnect within processor and memory chips. The evolution of monitoring and mitigation techniques rooted at the physical sources of heat. Hence, embedding temperature sensors in chips and managing the three main determinants of power consumption (voltage, clock frequency, and activity) at the chip and package level have been viewed as the immediate steps towards thermal management. The circle has been growing ever since. Cooling solutions also start at the nearest vicinity of the active devices in direct physical contact with the hot die using heat spreaders and heatsinks. Similarly, they evolve towards the upper layers of the computing stack.

There are interesting parallels to be drawn. Known methods at the macro scale have been adapted to micro and nano scales. Monitoring and cooling devices have been designed inspired by their bulk and discrete versions, finding novel embodiments as integrated (e.g., between tiers in a 3D IC, photolithographically patterned) and miniaturized (e.g., thin films, nano engineered materials) devices. In the reverse directions, the fundamental method of modulating a device's power and heat output via performance throttling has been migrated from the chip-level to the upper layers of the stack, including the memory sub-system, the communication network, and I/O.

In this chapter, major approaches at the system-level and above for mitigating thermal output in computing systems will be discussed. First, the role of the OS in coordinating thermal management will be described with highlights from various recent methods. Next, systems at a larger scale will be presented, including thermal mitigation in datacenters and supercomputers.

5.1 Mitigating heat dissipation at the OS level

The OS determines the task schedule and thereby directly impacts the hardware's activity. In most commercial systems three major layers coordinate power and thermal management: hardware (CPU), software drivers, and the OS. The industry standard ACPI is used in most systems to convey information from the hardware layer to the system layers and transmit control signals from the system to the hardware. Also, differences between specific policy implementations among different hardware architectures in implementing power and thermal management policies and data representation is resolved within this interface by translating the policies appropriately for the software drivers and the OS. An illustration of this hierarchy can be made using Intel processors as an example. At the highest level of abstraction there are Global System States dictating the mode of operation, such as active, sleeping, power down, shut off. These global states are further divided into the C-States (also called the processor power states), which provide multiple sub-state for the active and sleep global states. For instance, the active global state has four C-States associated with it, such as the highest power, fully active state C0 and idle state C1, and additional model-specific low power states with increasing latency to resume back to the C0 state. The C0 state in turn has a number of sub-states called the Performance States (P-States). These states correspond to various operation points of the hardware defined by a specific voltage and frequency.

ACPI establishes the main representations and data structures to enable the system software to track the state and actions of the hardware. However, the C-States and P-States can be implemented in various model-specific policies by the hardware. The BIOS and the software drivers are in charge of the necessary translation to facilitate seamless coordination between the ACPI and the underlying hardware policies. An overview of the relevant system components is illustrated in Figure 5.1.

Using the well-defined and widely accepted ACPI framework, most systems carry out dynamic thermal monitoring and thermal management. Various methods for sensing and policies executed by the hardware for performance throttling have been discussed in Chapter 2. In this chapter, we will turn our attention to policies and preventive methods that are applicable at the OS level and above for thermal management.

An important observation regarding thermal events and the role of the OS in influencing their evolution is that the rise and fall times of on-chip temperatures are typically longer than the decision intervals of the OS. This parameter depends on the thermal properties (thermal resistance and capacitance) of the hardware, which is a function of the materials used and the dimensions of blocks using each type of material. Although in some cases, it has been reported to be around 10 ms [1] for

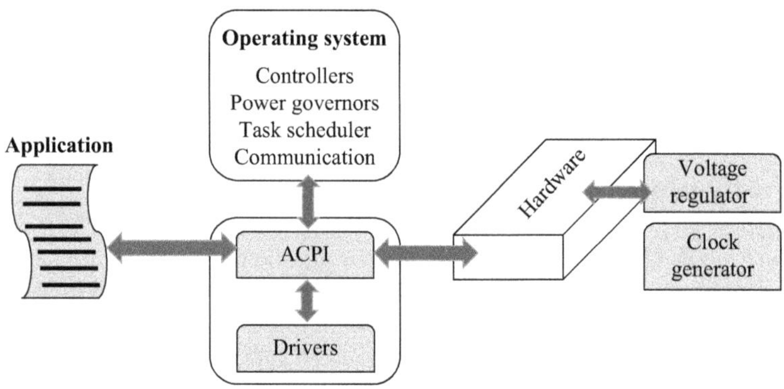

Figure 5.1 System-level view of the relationship between components that impact and/or manage thermal response

many modern architectures it is in the order of hundreds of milliseconds. For instance, for a POWER4-like architecture the thermal time constant has been reported to be around 200 ms [2]. On the other hand, schedulers and governor modules of the OSs operate with a sampling (tick) rate which is in the range of milliseconds. Hence, the OS has access to a relatively wide horizon of time to assess thermal state, make predictions, and perform high-level decisions to impact the thermal response in the longer time scale. Various run-time optimizations concerning the management actions of the OSs have been developed for this purpose. In the following, these thermal management methods will be discussed.

5.1.1 Thermal-aware optimizations

OS-level methods for thermal management focus on three main aspects: thread allocation and migration, task allocation and scheduling, and power management.

5.1.1.1 Thread allocation and thread migration

Thread migration is a technique originally employed for performance and load balancing optimizations, which later found uses in power, and thermal management in multicore processors as well. Also, it is used to exploit heterogeneous multicore chips by dynamically matching changing workload phases with the optimal choice of core. OSs are in charge of populating the run queues of cores and transferring the architectural state and data from one core to the destination core.

For instance, the Solaris OS utilizes a thread allocation method pursuing opportunities to utilize data locality and performs dynamic migration only if the load distribution is significantly skewed [3]. For each incoming thread, this mechanism first evaluates the amount of time that has passed since the last the same thread was assigned to a CPU. If that time period has not expired, the thread is assigned to the same CPU that it executed on most recently. This hopes to take advantage of the data that the thread could potentially use in the still warm cache of this CPU. If that time period has expired, the chances of the thread benefiting from the local data is reduced and the thread is assigned to a CPU that is currently running a thread with the lowest priority. This method also maintains a hierarchical grouping of threads, where each thread belongs to a local group, *lgroups*, and other groups are ordered according to a distance metric. In addition to this policy applied at the point of allocation, dynamic migration is invoked to balance the number of threads waiting to execute in the run queues of CPUs. Priority is given to balance the number of queued threads across physical chips first. This means, it is not desirable to have all cores in one chip to carry (however balanced) the computational load, while the cores in another chip are all idle. Instead, it is preferred to spread works across chips in a distributed system.

From a purely performance perspective, most OS policies aim at balancing the load and optimizing the rate progress in all threads. Therefore, these implementations of migration are oblivious to the thermal consequences of their decisions.

One of the early proposals to involve OS-level decisions in thermal management is the heat and run method [1]. An observation on the heat propagation dynamics inspired this method to manage multi-threaded applications, such that a mix of threads that activate a core's components maximally and simultaneously, so that the thermal headroom in a given core is maximally exploited, before migrating the computations to another destination. The main rationale is that regardless of whether only a sub-region of a core overheats due to elevated activity or majority of its sub-components exhibit thermal hotspots, the time required to cool of the core's peak temperature will be constant. Since, most heat dissipation occurs through the vertical thermal conduction paths towards the heatsink, lateral interaction matters little and all regions within a core can dissipate heat concurrently. Therefore, a thermally aware OS scheme assigns threads to cores such that there are threads in a given mix to utilize diverse components in a core (e.g., floating point and integer threads with complimentary usage of register files and issue queues). This thread assignment method is augmented with a migration scheme, which tracks the on-chip temperature sensors to detect when a core is approaching its thermal limit. Upon this trigger, the migration scheme moves threads from a hot core to another core with minimal thermal coupling and maintains uniform distribution of heat dissipation across the entire chip. An important parameter for the effectiveness of thermal-aware thread migration is the response time of the on-chip temperature sensors. For this particular method sensors are expected to consume a few microseconds to detect and report temperature. Among various alternatives, the policy, which identifies specific resources prone to create hotspots and aims at grouping threads with constrained total utilization of these resources to avoid hotspots, achieves best performance outcomes.

A heat balancer approach has been proposed using experiments on a dual-core Power5 processor [4]. While the default load balancing scheme in Linux would only intervene if the task queues of two processors are significantly skewed, this heat balancing scheme migrates a task with high predicted activity from an overheated core to a colder core even if the queues were perfectly balanced in terms of computational load.

An important parameter concerning thread migration is the frequency of migration. While re-allocation of activity around a multicore chip helps to uniformly distribute heat dissipation and temperature, migration cannot be invoked too frequently, since it incurs non-negligible performance overheads. Hence, a careful trade-off between thermal constraints and performance overheads should be pursued. One study investigated a limited scope of pingponging activity between two sites, which minimizes the migration overheads [5]. Additional benefits of reducing power density and temperature on leakage power has been quantified as part of this study. Different architectural configurations have been evaluated, where two symmetric sites for migration could be exact replicas in all resources of a core, or they can have split execution units and share other substructures in various combinations (e.g., register file, data cache, instruction cache, issue queue, rename table). Another proposal pursued a novel microarchitecture that is inherently more suitable for thread swapping with the notion of helper engines [6]. The helper engines are specially designed structures to buffer thread states for low overhead migration. Large area consuming resources are shared across threads with minimal duplication for smaller microcore structures. The overall architecture uses two identical cores (main core and extra core) and when migration is triggered by an on-chip sensor on one of the cores, then computation is moved to the cooler core until it heats above the critical threshold and another swap is required.

Another study of thread migration under a temperature constraint has tackled the performance overhead of migration in a more direct manner without re-design of the core [7]. The total number of threads that are allowed to execute on a multicore chip has been constrained by the total power consumption allowed, which in turn is linked with temperature. The problem of identifying the best set of threads to execute at a given time is formulated with the thermal fairness concept, which dictates that threads with similar activity features should be allocated with similar shares of power budgets and each thread should be guaranteed a minimum performance (power budget) at all times. The OS decides which threads should be executing at a given time and for how long. The idea is to embed the abovementioned thermal fairness into the priority function of the OS.

In this case, the thread migration is not intended as a solution to directly help cooling in the system. A dynamic performance throttling method similar to the scheme used in TM1 method of Intel processors has been used in this study as the primary mechanism for temperature control. Global clock gating is applied upon a core's sensor triggering a thermal emergency. The function of the thermal-aware thread migration is to help the system maximize its performance while DTM is activated. Specifically, this is achieved by migrating threads to those cores that have not saturated thermally, in an attempt to maximize the fraction of execution time where majority of the cores are operating near the saturation point. The thermal saturation point is defined as the temperature limit slightly below the threshold for a sensor to trigger throttling action. That would correspond to the operating point allowing the highest power consumption without incurring performance penalties due to throttling. The method categorizes threads into two types, saturating and de-saturating. Saturating threads would generate heat output to exceed the safe temperature threshold within the period of t_{on}. De-saturating threads incur a heat output that would remain below this limit. For a given thread C, the duration of time slot and power budge allocated for it is then determined by inspecting the other threads that are simultaneously present in the system. If the rest of the threads are all similar in type to C, for example, all are saturating, then thread C's allocation of execution period would be scaled down. Whereas, if all other simultaneous threads are de-saturating, then C would be allocated more execution time and power.

A method, which incorporates migration overhead into the decisions-making mechanisms, performed a thorough evaluation of the overheads associated with the involvement of the software layers (OS, middleware, and communication libraries) in facilitating migration [8]. Thereby, cases where the benefits of migration in thermal balancing will not be overshadowed by the performance overheads. In this particular study this is an even more pressing issue, since a MPSoC for streaming applications is considered where the quality of service (QoS) is of utmost priority. The extent of effort that needs to be placed towards thermal balancing of individual processors in the MPSoC has been defined with a temperature interval around a specified mean temperature for each processor. The rationale is that if the thermal-aware load balancer can ensure to maintain the temperature of each processor with this range, consequently, the overall system will be thermally balanced. Using on-chip temperature sensor feedback, a migration trigger is activated, when the temperature of a processor reaches either its upper or lower limit of the thermal interval. In these cases tasks are moved away (towards a processor at the lower end of its temperature limit) or into (from a processor at the higher limit of its temperature limit) the run queue of this processor, respectively. For each migration action, one source and one single destination processors are selected according to a set of criteria concerning the thermal states of the processors. The next step is to determine the exact amount of workload to be migrated from the source to the destination. It is assumed that tasks might be migrated between this pair of processors back and forth at a frequency that is proportional to the initial temperature difference between them, until a balance between them is achieved. The dynamic evaluation of the migration cost takes this frequency into account scaled by the amount of data that needs to be transferred for a set of tasks. A practical realization of this thermal balancing migration scheme constructs its migration cost function according to the impact of the migration overhead for a given underlying software infrastructure and the tolerance in the QoS metric for a given application.

Thread migration schemes applied at emerging technologies are also drawing attention. One proposal aimed at thermal management of a stacked DRAM-core IC with thread migration integrated into the OS context switching mechanism [9]. One of the proposed schemes rotates threads among cores in a round robin fashion at a fixed period of 10 ms. Another scheme utilizes temperature sensor feedback to swap tasks between the hottest and coldest core, provided that the temperature difference between these two cores is larger than a pre-defined threshold.

A comprehensive study presented detailed thermal measurements of a commercial processor test chip particularly emphasizing the thermal variation across and within cores [10]. Core hopping, that is, activity migration across cores, has been analyzed as a means of thermal management. It is argued that a thorough thermal characterization of the underlying hardware should guide the OS-level migration and scheduling policies. Preventive activity migration by setting asymmetrical intervals for the intrinsically hotter and colder cores has been shown to be effective.

A recent task migration method emphasizes the physical aspect of re-arranging workloads of a core in relation to the physically adjacent cores [11]. Migration is triggered upon a temperature sensor reporting a core's temperature exceeding a threshold. The selection of a destination core to migrate threads now considers the thermal impact of the increased activity at the destination onto the thermally coupled neighboring cores.

Finally, we note that thread assignment and migration can also be performed by hardware mechanisms, however, we will not elaborate in detail on hardware-based schemes since they are not within the scope of this chapter [12].

5.1.1.2 Predictive thermal management at the OS level

OS-level methods for thermal management fall into one of two main categories: reactive and predictive. Reactive methods utilize on-chip temperature sensor feedback only, such as the thread migration schemes discussed earlier. Predictive methods may also utilize temperature sensor feedback, however, they often combine this information with other performance indicators (e.g., performance counters, system state) to predict the future thermal state and perform preventive measures to avoid thermal emergencies and performance throttling. These schemes employ a run-time predictor for the peak core temperatures and utilize this information to perform OS-policies. In the following, we will review existing approaches to load balancing and thread migration using predictive schemes.

A HY predictor has been proposed to combine a thermal model reflecting application- and core-specific features [13]. Based on the feedback from this predictor, the OS performs thread migration to move computation from the hottest predicted core to the coldest core. Also, the priority scheduling is adjusted to modulate the chip resources for power density control. This new OS policy has been embedded into Linux running on an Intel Quad-core system. The application-specific component of the predictor aims to capture the variations in thermal response within shorter time intervals following the application's fluctuation of activity. The RLSM has been used to generate a regression model to predict future application-specific temperature. Such a regression model essentially relates a set of system parameters to the system's output (temperature in this case) with a linear expression as follows:

$$y = \alpha_1 u_1 + \alpha_2 u_2 + \cdots + \alpha_n u_n \tag{5.1}$$

$\boldsymbol{u} = [u_1, \ldots, u_n]$ is the vector representing the system parameters in the form of an input vector for the regression model. For instance, an input vector might be comprised of number of instructions, temperature reading, and cache misses and the observed output is temperature at the end of a sampling interval. The parameters $\alpha_1, \ldots, \alpha_n$ need to be estimated to construct the model from a given set of observed input vectors and system output. An estimator for the α coefficients is used to derive them for each prediction interval $k+1$, based on the inputs and outputs observed up to the kth interval and the coefficient values estimated for the previous intervals. The core-specific temperature prediction is based on a computational model of temperature relating the rate of temperature change with a constant coefficient to the temperature difference between the current temperature and the steady-state temperature. This first-order differential equation is solved at the boundaries of ambient temperature at $t=0$ and steady-state temperature at $t=\infty$. The constant coefficient is derived empirically based on the rate of heating and cooling of a given processor under 100% workload. Steady-state temperature under full utilization is computed offline with thermal simulation. The resulting predictor for core-specific temperature is as follows:

$$T(t) = T_{SteadyState} - \left(T_{SteadyState} - T_{Amb}\right)e^{-bt} \tag{5.2}$$

At the start of a prediction interval, temperature is predicted using a weighted sum of the application- and core-specific predictions. This predictor is embedded into the standard Linux scheduler. Migration is initiated in a two-step process. The core-specific predictor is first used to predict temperature into a farther point in future time. If this prediction exceeds a pre-set trigger limit, then the application-specific short-term predictor estimates the future temperature and the time when migration should commence once the predicted temperature exceeds a second threshold justifying the migration overhead.

Another popular tool for prediction is the ARMA method. A method to proactively allocate workload in anticipation of thermal emergencies and aiming to avoid those has been developed using this paradigm [14]. Furthermore, this method is enhanced with dynamic adaptation so that variation in workload characteristics can be sensed and the predictor can be updated at run-time. The ultimate goal is to utilize this predictor to migrate tasks to balance the temperature distribution in a multicore chip.

The ARMA model captures a discrete time series of values x_t. In this model, a value x_t at a time instant t is assumed to correlate with p past values in the series through an AR coefficient a and with q white noise error terms with moving average coefficients b. This model is called to be order (p, q) and be expressed generally as:

$$x_t = \sum_{i=1}^{p} a_i x_{t-i} + \sum_{i=1}^{q} b_i e_{t-i} \tag{5.3}$$

The noise (also called the residual) terms are assumed to be uncorrelated with a normal random distribution, $e_t \sim N(0, \sigma^2)$.

The process for generating an ARMA model is illustrated in Figure 5.2. The workload allocation scheme utilizes specific methods to realize these steps [14].

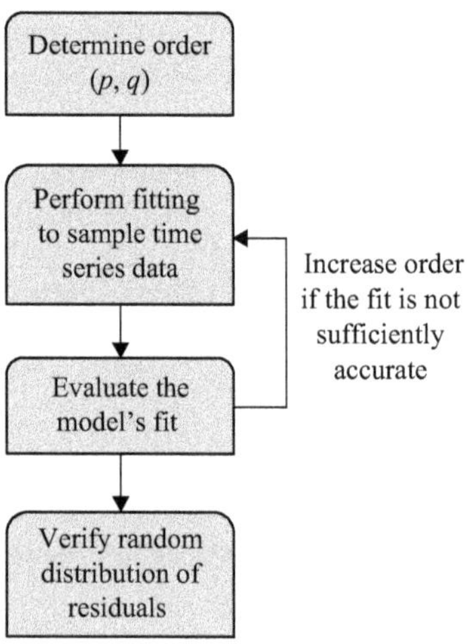

Figure 5.2 The process flow for generating an ARMA model

Generally, it is desirable to generate a model with acceptable accuracy with the smallest possible p and q. This has been accomplished in this particular approach by starting with the smallest sized model ($p = q = 1$) and perform an initial fit using the popular least squares fit method. The accuracy of the resulting fit is then evaluated and the order is increased until the fit passes the accuracy test. Expected accuracy of the fit can be evaluated in various ways. For example, the square of the correlation factor between the output and the inputs of the regression helps to explain the fraction of the variation in the output that can be directly attributed to the variation in the least squares regression. In this specific study, the usefulness of the fit is measured with an alternate method using the concept of final prediction error. This metric relates the usefulness of the fit to the size of the time series (N), the order of the ARMA model ($n = p + q$), and the variance of the noise term (V) in the model as follows:

$$\textit{Final prediction error} = \frac{1 + \frac{n}{N}}{1 - \frac{n}{N}} \times V \tag{5.4}$$

An iterative loop is executed as shown in Figure 5.2 until the final prediction error drops below a threshold.

At the last step, a final property of the ARMA model is verified, which is the assumption of randomly distributed noise terms. This particular modeling method accomplishes this using the autocorrelation function. This function can be used to verify that in truly random distribution of the noise terms, the autocorrelation is close to zero. A limited confidence interval has been defined in this study and the ARMA model passes the verification step if the autocorrelation factor for the majority of the noise series falls within this interval.

Using this process, a predictor for a core's temperature has been derived from temperature measurements as the observed data trace. In a multicore chip, one predictor for each individual core is generated to capture individual thermal behavior. An enhancement to the prediction method has been introduced through online adaptation. Such adaptation is necessary due to the varying nature of workloads. Hence, the temperature time series of a system will only be stationary within a specific workload phase, while the thermal behavior of a different phase might largely diverge from the data used to train the ARMA model. It would incur significant overheads to build and train ARMA models for all possible workload phases. In order to mitigate this issue, continuous validation of the ARMA model has been proposed [14]. This online validation step involves detection of whether the noise terms deviate from the distribution characterized at the training phase. A statistical tool has been used to quantify the amplitude of deviation of the mean of the noise terms from zero. A threshold on this deviation is applied that is in the order of $3 \times \sqrt{V}$, using the variance of the noise terms in the training data as a reference. Upon detecting such a deviation, a new ARMA model is computed at run-time.

As part of this same study, other alternative predictors have been evaluated, such as the exponential averaging predictor, a branch predictor-like history predictor, and a recursive least squares predictor. The default policy mechanism in the Solaris OS (DynamicLoadBalancing) has been enhanced with these predictors. The ARMA model uses temperature measurements reported by sensors from each core. The predictors are used to implement a proactive migration policy, which identifies a CPU that is predicted to be hot in the future and move threads away from it. Also, the allocation decisions are revised dynamically using the temperature predictor. Initially allocation is performed using the locality principle described earlier. Then, at each decision epoch, if a core is identified as likely to become hot by the next epoch, threads are moved from its run queue to an alternate core. Moving waiting threads is a more cost-effective policy compared to migrating a running thread, which incurs performance penalties. Also, the number of migration has been bound to control the performance overhead for the general case.

5.1.1.3 Task allocation and scheduling

Another critical decision made by the OS that affects the thermal profile is the allocation and scheduling of tasks. As it is the case with all design and operations problems, planning decisions made at the earlier stages of the process have the largest global impact on the system. OS-level decisions are crucial for this very same reason. The OS has access to a wide variety of information regarding tasks, such as deadlines, the expectations regarding responsive of the system from the user's side, memory access patterns, data types, and I/O characteristics. In that sense, the OS is capable of making much more impactful and well-informed decisions.

Modern OS schedulers prioritize fairness among tasks in their decision-making mechanisms. The amount of time allowed for each task to use a processor is maintained with a balance. For example, in the Linux 2.6 OS, each task is associated with a virtual time. The priority of a task is linked to a decay factor, which continuously scales the time allowed for a task to use a processor. Lower priority tasks have larger decay factors in comparison to higher priority tasks. A time ordered red-black tree is used to maintain tasks ready to run according to the amount of virtual time they possess. Those with the smallest amount of time are picked first to schedule next following the fairness principle. Naturally, organizing the processor activity purely from a fairness point of view can create thermal imbalance, since different tasks may incur a different amount of dynamic power consumption as well as they may activate the local execution units within processors in an unbalanced manner. Various academic studies have, hence, tackled the problem of enabling thermal awareness in task scheduling and allocation.

One of the earliest studies on the thermal impact of task scheduling focused on thread assignment policies to test the limits of existing policies [2]. The analysis was performed with thermal simulations on a Power4-like architecture. A number of policies are tested with a 10 ms decision interval similar to the default period used by the latest version of Linux OS at the time of this study. Threads are executed in the first time slice and the processor's temperature is collected for each thread as a representative indicator of their heating behavior. This analysis found that for different invocations of the same thread from workloads of the SPEC2000 benchmarks, the variation of peak temperature was confined within 5°C. Hence, it has been deemed safe to derive the thermal behavior of a thread based on this initial test run.

The metric in the evaluation is the success of each scheduling policy in maintaining the chip's temperature below the safe threshold, set at 85°C. The policies tested were random policy, average temperature policy, maximum temperature policy, and minimum temperature policy. The random policy assigns priorities to all threads fairly and picks the next thread to be scheduled randomly. Hence, it is completely unaware of thermal implications. The fact that threads are randomly selected actually helps peak temperatures in critical execution blocks such as the floating point unit to taper off. This is because a floating point thread that makes heavy use of this block causing its temperature to rise might be followed by an integer computation workload that does not utilize this block, leading to an eventual stabilization of the block's temperature. While uncontrolled rise of temperature is thereby avoided, this does not prevent high temperatures above the pre-set threshold. It has been found that the chip spends 96% of execution time above the safe threshold.

The average temperature policy associates each thread with the average temperature observed through its execution during one time slice. At run-time, when the chip temperature exceeds the threshold, the scheduler prioritizes threads with the smallest average temperature. While this policy seems to have improved the thermal response, variation in temperature in a thread's lifetime still causes the temperature of some blocks to continuously rise. The maximum temperature policy primarily aims to achieve peak performance and schedules threads with higher temperature characteristics in sequence, leading to violation of the thermal threshold in 98% of the execution time. Finally, the minimum temperature policy manages threads with priority on temperature whenever the threshold is exceeded. For each decision interval, the processor block with highest temperature is identified. Then, from the thread queue, the thread that had incurred the least temperature rise on this block in its test run is selected to execute next. When the chip's temperature is below the threshold, this scheduler resumes to maintaining fairness among threads and allows execution of high-temperature threads to take advantage of the low-temperature periods. This policy succeeds in keeping the peak temperature below the threshold for 99% of the execution time. This study helped to demonstrate the importance of scheduling policies in shaping the thermal map of a processor and the extent of variation among different policies. Also, it confirms the importance of on-chip sensors in guiding these policies, since the thermal metric of each thread in a real system needs to be derived from actual measurements.

A follow up study evaluated the next generation Power5 processor architecture with live experiments and provided thermal-aware solutions at the system software level [4]. This study is motivated by the observation that the unit-level resource usage of a thread is a strong indicator of its thermal impact, since unutilized blocks are placed in low power state through fine-grain clock gating. This heavily influences where hotspots will occur. The OS's access to knowledge of dynamic thread priority and deadlines for real-time tasks is exploited to dynamically re-arrange task execution and thereby the spatial variation in power density towards a smoother distribution. Twenty four digital sensors on the Power5 chip are sampled every 4 ms (corresponding to the schedulers tick) to track temperature. One of the approaches tested in this study was to allow tasks visit a switch between the two cores of the dual-core Power5 chip at every scheduling tick. This approach has been characterized as one leveraging the spatial heat slack. The complimentary approach aims to target the temporal heat slack. When a mix of workloads exhibits varying degrees of thermal aggressiveness, it has been suggested to interleave hot and cold threads over time to spread the heat dissipation across a longer timescale. This yields reduction of peak temperatures.

A thermal-aware scheduler aiming to preserve the fundamental scheduling principles of existing OSs has been proposed to adapt the Linux 2.6 kernel on a multicore processor [15]. This affinity-driven scheduling objective of the default Linux scheduler may cause a thread to execute on the same CPU repeatedly to exploit the data in the warm cache and the load balancer may not intervene for a long period of time or not at all, unless a significant imbalance among CPUs' task queues is observed. This study points to the potential hazard of this scheme to allow a thermally intensive task to hit the same CPU repeatedly causing steep hotspots. A thermally aware alternative is developed based on the well-known round robin scheme. The basic round robin scheme allocates an equal time slice for each thread in the queue. Threads are pulled from the queue in order and allowed to use their time slice on whichever core they happen to be assigned to. This scheme is altered with thermal considerations as follows. The system tracks die temperatures using the on-chip DTSs. When the die temperature exceeds the system's safety threshold, threads are scheduled to spend an equal fraction of their time on each processor to spread the heat dissipation evenly across all cores. This is expected to lower the peak temperature of the hottest core. For a 2-core chip tasks are migrating between the two cores periodically. For a quad-core chip tasks are rotated to spend equal amount of time on each core.

Another thermal-aware scheduler was proposed to modulate time allocation priorities to processes when a thermal emergency demands immediate attention [16]. Each process is associated with a thermal intensity using profiling information from test runs of the process and collecting performance counter statistics. The profiler implemented as a shared library collects the frequency of accesses made by a process to the processor resources, which correlate with power consumption and the resulting power density. Thermal-aware intervention by the OS occurs under the following circumstance. When the on-chip temperature sensors indicate that a thermal threshold has been exceeded, the OS takes action by downgrading the priorities of all processes that classify as having high thermal intensity according to the profiling statistics. The priorities are decreased gradually in increments to minimize the performance impact and to be able to recover quickly as soon as the thermal emergency subsides. Nevertheless, the increments are set to a significant enough amount to overwrite any balancing that may occur for fairness at this point in time. Thereby, hot processes receive a shorter slice of time for processor usage compared to colder processes. The rationale here is to still maintain a close to full utilization of the system and accomplish that by running the colder processes. When the on-chip sensors report the chip temperature to have dropped below the critical threshold, priorities of all processes are restored. At the other extreme, reduction of priorities continues until all present processes have reached their lowest possible priority. If the system does not recover from the thermal emergency despite all processes having their lowest priorities, all priorities are restored and the schedule is re-created in an attempt to find a superior solution.

The trade-off between conforming to a thermal constraint while maximizing throughput has been investigated as part of task allocation for multicore processors [17]. An online task to core assignment algorithm is developed that can execute within a 5 ms decision interval that is typically sufficient for OSs for a large number of cores (up to 128). The problem that is tackled in this method is to assign a number of tasks (n_t) to n cores and also determine their clock speeds such that the temperatures of the cores do not exceed T_{max}. The optimization objective is to maximize the throughput for the system defined as:

$$Throughput = \frac{1}{t_s}\int_{0}^{t_s} \boldsymbol{w}^T s(t)dt \tag{5.5}$$

$\boldsymbol{w}$ is a weight vector, t_s is the decision interval for task allocation, and $\boldsymbol{s}(t)$ is the vector of speed assignments to cores at time t. The computational complexity of this optimization problem is improved by making a simplifying assumption on the objective function. Instead of treating throughput as a continuous entity, the instantaneous throughput at the start of each allocation decision interval is maximized. This simplifies the task of determining the speeds of the cores and with a given speed assignment, it helps the method to represent the subsequent task assignment problem as a linear optimization problem. A classical optimization technique from the OSs domain called the Munkres algorithm [18] is then utilized to solve the task assignment problem in real-time.

In addition to CMPs, MPSoCs have also been subject to various studies on thermal mitigation. A static task scheduler aiming to minimize the peak temperatures and smooth out thermal gradients for a MPSoC has been proposed using the ILP paradigm [19]. The scheduler aims to meet real-time task deadlines and inter-task dependence constraints while minimizing the temperatures of hotspots and the magnitude of the thermal gradient across the chip. A task graph with known dependencies, task deadlines, and worst-case execution times (WCETs) for each task is assumed to be presented as input to the scheduler. The objective function of the ILP formulation has two components. The first component incorporates the impact of temperature with a quantity, which denotes the amount of time spent by a processor above the safe thermal threshold. A summation of all such times across all processors along with the decision variable attaching a certain task to a processor is computed. The second term signifies the amount of overlapping activity in terms of tasks with overlapping execution times across two physically adjacent processors. Minimizing this intensive period of high activity helps to reduce hotspots. The time spent above the safe thermal limit is first approximated coarsely by the WCET of each task and an initial task assignment to processors is found solving the ILP comprised of the objective function, dependency constraints, and task deadline constraints. The resulting task allocation is ported into a thermal simulator to derive the actual peak temperatures of the processors. These temperatures are then used to refine the estimation of the time spent above the thermal threshold and the ILP is solved again for the final task allocation. An enhancement to the scheduler is also proposed to relax the assumption of knowing the exact structure of the task graph a priori. This may not be applicable in practice. The modified HY scheduler leverages a dynamic scheduler to make run-time adjustments to the scheduling decisions using temperature sensor feedback [20].

This dynamic scheduler has been equipped with a few modes of varying complexity. The simplest approach attempted was to perform scheduling reactively based on temperature sensor readings. At the start of each scheduling interval, the incoming tasks are scheduled on the coldest processor. An immediate variation to this scheme was to pick a cold processor with the largest number of idle adjacent neighbors, in an attempt to minimize the lateral heat transfer across processors. The third policy aims to find a middle ground between load balancing and temperature optimization. Each processor is annotated with a probability of having the next ready task scheduled on it. These probabilities are updated at the start of each scheduling epoch taking the temperature history of the processors into account. These probabilities are increased or decreased until they saturate at 0 or 1. At one of the extreme case, if a processor has exceeded the safe thermal threshold T_{max} of the chip in the previous scheduling interval, its probability is set to 0 for the immediate interval. A second temperature limit T_{low} has been defined as an intermediate state indicating a processor approaching the peak. T_{low} and T_{max} have been set to 75°C and 85°C, respectively. At the other extreme, if a processor's temperature remained between T_{low} and T_{max} during the most recent interval, its probability is not modified. For all other processors that did not exhibit a peak temperature above T_{low}, the tuning factor is determined as follows [20]:

$$W = \frac{\beta}{\frac{T_{avr}}{T_{max}}} \tag{5.6}$$

The thermal interaction of on-chip elements goes beyond thermal coupling between cores. The impact of temperature on the network of chip infrastructure has been studied in this context with a reinforcement learning method for thermal-aware task allocation [21]. The optimization objective is to minimize the global peak temperature of the chip. Task allocation to cores affects the data traffic and the resulting thermal profile of the network on chip routers. Therefore, these components also need to be taken into consideration. This specific method proposed in this study uses reinforcement-based learning to make the task allocation decisions according to some prediction of the subsequent thermal state. Then, using actual temperature sensor feedback these decisions are evaluated and the reward metrics of those actions that led to better solutions are continuously refined. The reward metric of this reinforcement learning instant has been set as:

$$Reward = T_{lim} - T_{peak}^{observed} \tag{5.7}$$

At each time step t, a utility function Q maps a given reward metric r, task allocation decision a, and thermal state *tem* collected from sensors, to the expected temperature reward in the system. This utility function combines the expectation over the entire horizon from time step t into infinity, while future predictions of rewards at steps $t + 1$ and onward are downscaled by a discount factor γ. Q helps to construct a policy for task assignment as follows [11]:

$$Q(tem, a) = E\left[\sum_{i=0}^{\infty} \left(\gamma^i r_{t+i} | tem_t = tem, a_t = a\right)\right] \tag{5.8}$$

At each time step $t + 1$, having access to the history of the allocation decision a_t at time t and the temperature outcomes tem_t and tem_{t+1} at the moment, the generalized utility function shown in (5.8), is approximated as:

$$Q_{t+1}(tem_{t+1}, a) = Q_t(tem_t, a) + \alpha(r_t(tem, a) + \gamma \max_{a'} Q_t(s_{t+1}, a') - Q_t(s_t, a)) \tag{5.9}$$

This iterative equation approximates the utility by using a history window of one time step and introduces the so-called learning rate parameter α to facilitate convergence of the iterative model. After each task allocation, thermal state is collected from on-chip sensors. Using this information the reward of the action is evaluated and the utility function is refined.

OSs deployed in 3D IC implementations of CMPs and MPSoCs are in need of better awareness of thermal management. In response to this growing need various techniques were developed. An important optimization parameter is the extent to which tasks that incur large heat dissipation are prevented from running in adjacent cores along the vertical direction. This is one of the main causes of significant heat being trapped within the inner tiers of a 3D IC and creation of hotspots that jeopardize reliability and increase leakage power.

A method developed to mitigate formation of hotspots due to vertical stacking of high activity tasks introduces two concepts: supercore and supertask [22]. All cores that are vertically aligned across all tiers are grouped into one supercore. Since the activity on these cores will intimately affect each other's thermal profile, task scheduling decisions within this supercore needs to be made in a coordinated manner. A supertask is a collection of tasks that are assigned to one supercore. One important distinction observed by the task scheduling algorithm is the customization of the scheduling algorithm according to the specifics of the 3D IC's floorplan. There are two main approaches to laying out major blocks such as cores and caches. The first approach is to layer them homogeneously, repeating the same basic pattern in all tiers. This simplifies the placement of TSVs and hence, achieves a fixed pattern for the physical design of the communication interface. The disadvantage of this floorplan is that blocks with similar thermal profiles, for example, high power density cores, will be perfectly aligned and congregated hotspots will be even more profound. The second approach is to modify this regular pattern according to thermal interactions and avoid direct overlap between high power density cores between the

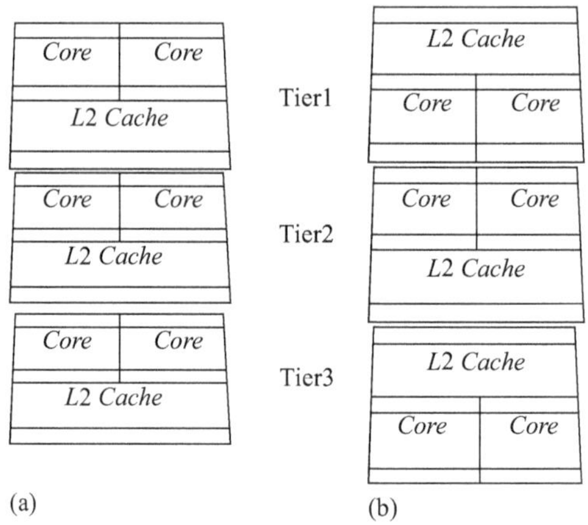

Figure 5.3 Representative floorplans for 3D CMPs: (a) A homogeneous floorplans where cores are aligned along the vertical direction and (b) an irregular floorplan, where the layout of two consecutive tiers are symmetrical with 180° rotation, so that cores do not overlap

immediately adjacent tiers. This is usually done by designing one main pattern, where high power density and lower power density (e.g., caches) blocks are interleaved and then rotating this pattern by 90° at each tier. Thereby, in two consecutive tiers a core and a cache block mostly overlaps vertically. Figure 5.3 illustrates examples of homogeneous and irregular floorplans for an abstract IC using two cores per tier. As mentioned earlier, the interconnect structure for this design is more challenging, since it is harder to extend TSVs through the interiors of cache blocks with dense and homogenous layouts of cell arrays. Nevertheless, in the face of stringent thermal challenges, this is an overhead that might be allowed in the design.

The scheduling algorithm managing supercores and supertasks for the regular floorplans performs a two-step optimization. For a 3D IC comprised on k tiers supertasks are formed from k tasks, since there will be k cores along one vertical column. Out of m available tasks, k supertasks are formed such that the total power consumptions across the supertasks are balanced. A simple bin packing heuristic suitable for real-time execution is used, where m tasks are first sorted in increasing order of their individual estimated power consumption. Then, following this sorted order each task is placed into a supertask bin with the least current total power consumption. Assigning such power-balanced supercores is expected to also generate uniform power densities across supercores. At the second step of the scheduling algorithm, at any scheduling epoch (every 8 ms), the temperatures of all cores within one supercore are added to obtain a representative thermal metric for a supercore. The supercore with the highest metric is assigned with the supertask of smallest power consumption. Similarly, within a supercore, a supertask is then broken into individual tasks for per-core assignment, where the core with highest reported temperature receives the tasks with smallest power consumption from the bundle.

For irregular floorplans, the scheduling algorithm differs as follows. The tasks are binned in three different ways, to create a varying level of diversity in their power profiles. The effective thermal resistance from each irregular supercore to the heatsink will not be identical as in the regular case. If the current temperature difference between supercores is large, that means a significant temperature adjustment will be neceassary during the next scheduling interval. Then, supertasks with large variation among them will be selected and a skewed power assignment to the supercores will be allowed. If the temperature difference between supercores is small, supertasks can still be formed in a balanced manner.

A follow up work combines dynamic voltage scaling and task scheduling for 3D ICs and similarly incorporates the important notion of thermal correlation along the vertical dimension to optimize voltage and task assignment decisions [23]. This method utilizes the supercore concept introduced earlier [22]. Similarly, vertically aligned cores are grouped into a supercore. For instance, in a 2-tier IC with four cores in each tier, two vertically neighboring cores form a supercore and the chip is thereby divided into four supercores. At the start of each scheduling interval of the OS, ready tasks are divided into groups of two (for a 2-tier IC), such that the tasks with larger predicted power consumption are paired with tasks with smallest predicted power consumption. The scheduler aims to assign the task with the smallest power consumption to the hottest core at the current time in order not to exacerbate its thermal load.

Another study first starts out investigating the effectiveness of existing OS policies for thermal management and moves on to propose a thermal-aware job scheduler [24]. This study correlates the thermal stress of a core with its distance to the heatsink and assigns a thermal index $Index_c^{Therm}$ to each core accordingly. Cores that are located in the upper tiers closer to the heatsink have access to better heat removal. A probabilistic scheduler is employed, where during each scheduling interval i, each core c, has a probability P_i^c of having a new task assigned to it. This probability is adjusted at each interval with a positive or negative increment W based on the following criteria:

$$if\ \Delta T = T_{lim} - T_{avr}^c \geq 0 \quad then\ W = \beta_{inc} \times \Delta T \times \frac{1}{Index_c^{Therm}}$$

$$if\ \Delta T = T_{lim} - T_{avr}^c < 0 \quad then\ W = \beta_{dec} \times \Delta T \times Index_c^{Therm}$$

T_{lim} is the thermal threshold imposed on the cores by the power and thermal management mechanisms. T_{avr}^c is the historical average temperature of core c. In this particular study, temperatures reported for the last 10 scheduling intervals have been used. Finally, β_{inc} and β_{dec} are empirical constants.

5.2 OS-level thermal policies for embedded and real-time systems

A major domain in computing hardware is represented by embedded systems. They are primarily characterized by the real-time nature of task management. Other high-performance computing systems do also serve real-time applications with unique performance expectations. In this section, a separate discussion will be devoted to the embedded systems in particular and real-time systems in general. Their run-time systems are also being enhanced with various methods towards more effective management of heat dissipation and temperature control. Compute intensive applications that are typically executed in embedded and real-time systems, such as signal processing, video encoding and decoding can elevate operating temperatures. Especially considering the fact that these systems may be limited in form factor and cannot host heavy duty cooling solutions, the thermal profiles can rise quickly.

Embedded systems can be configured in two major ways. The first is to generate task allocation and scheduling statically for a specific task graph, which is expected to repeat periodically on the embedded system. An individual task is often characterized by its WCET. The embedded system is typically represented by one or multiple processing elements (PEs) that are capable of executing the nodes in this task graph. Figure 5.4 summarizes major properties that define a task graph targeting an embedded system. Figure 5.5 illustrates an example of execution for a periodic task graph on an embedded system with two PEs.

Even static allocation and scheduling decisions can be performed with thermal awareness. Techniques that fall into this category often perform allocation and scheduling in multiple iterations (e.g., list scheduling, force-directed scheduling), where in each iteration tasks with highest criticality are processed. Once decisions are fixed for a subset of tasks, the criticality of the remaining tasks are recomputed to start the next iteration. These loops continue, until the entire task graph has been processed. The criticality of tasks on these methods are primarily determined

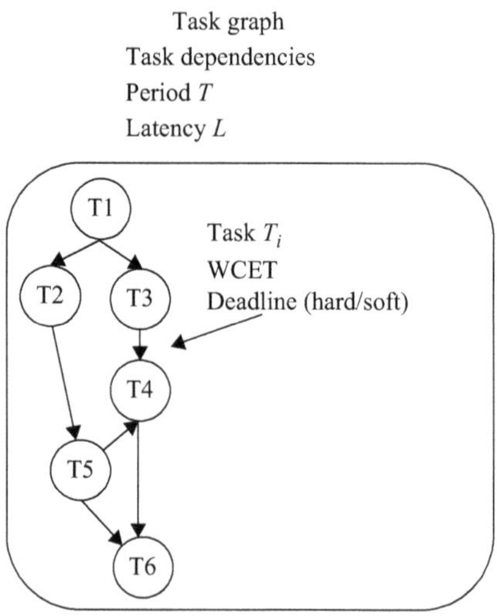

Figure 5.4 A task graph with dependent tasks. Each task is characterized with a WCET. The task graph can be associated with a period, where tasks arrive at a period T

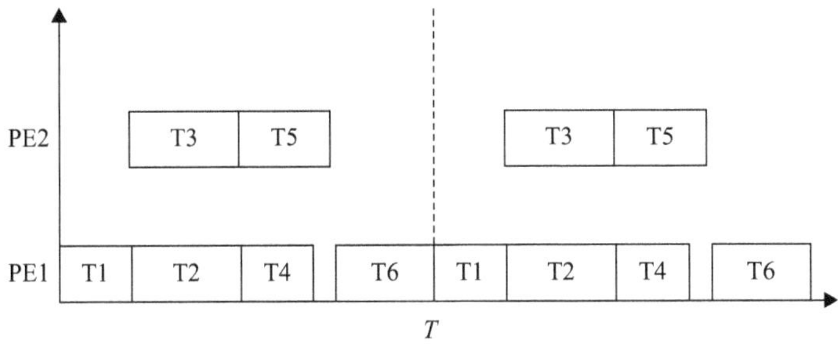

Figure 5.5 A periodic task graph scheduled on a system with two PEs. The PEs can be CPUs, ASICs, etc.

according to their contribution to the total critical path and/or throughput of the final schedule and also in terms of the slack between the current time step and the deadline of the task (if it is defined). A thermal-aware method to extend this basic principle has been proposed, where this criticality metric is augmented with the expected peak temperature of the entire chip using a thermal simulator in the inner loop [25]. Naturally, the next logical extension is to consider run-time scheduling mechanisms. A heuristic for thermal-aware dynamic scheduling of tasks on embedded MPSoCs has been proposed by the same study [25]. This algorithm identifies a task that can execute on an alternate PE (in this case assuming the system is comprised of homogeneous PEs) and repeatedly re-schedules this task on alternating PEs. While specific details on this algorithm are not provided in this study, various alternatives can be envisioned. The basic rationale behind the method is to maintain a balance of task loads across multiple PEs, so that their effective power density is uniform. Assuming the PEs are identical, it is argued that the total power consumption will remain unaltered regardless of where the migrating task executes. While this optimization can be achieved by migrating a single task back and forth, it can also be extended to identifying a pair of tasks where one has more time slack. By hosting the task with more slack, each PE might benefit from that downtime to cool down during the idle time. Furthermore, during that idle time the PE can be placed in a sleep mode to further reduce power consumption.

A new scheduling policy has been proposed for thermal-aware load balancing and scheduling placing special consideration on the physical aspects such as the location of different PEs and their thermal history collected from on-chip sensors [26]. Existing policies that have been embedded into traditional OSs such as random scheduling (shuffling tasks among cores), assigning the next task in the queue to the core with lowest current reported temperature, and assigning a task considering the temperature of adjacent cores as well, have been acknowledged in this study and the policy has been developed as an extension on these techniques. The proposed policy extends the notion of utilizing core temperature by incorporating past history of a neighborhood of cores. Also, this policy applies to aperiodic task graphs where tasks have soft deadlines and they can also be pre-empted. The quality of a schedule is measured by a cost function with two terms. The first term denotes the difference between the finish time of each task t_j and its soft deadline. If the deadline has been exceeded, this factor into the cost function as a penalty proportional to the execution time of each task. The second term indicates the thermal impact of assigning a task t_j onto a processor, given the current thermal state of the processor and the projected thermal state after t_j completes execution. The projection is performed by using the average temperature change; this task has incurred on the system in its history over a fixed period of time. At the same time, a task is not allowed to execute on a processor if it is found that the projected temperature will exceed the safe thermal threshold defined for the hardware. If no task in the ready queue can execute on a given processor without causing it to exceed its thermal threshold, then, that processor is given idle cycles for a few scheduling intervals. Another aspect of this policy is that it also incorporates the notion of thermal saturation for cores. The policy avoids having a few cores that have reached their allowed peak while other cores are significantly colder. Task migration is initiated on cores that have saturated to offload their tasks onto the historically coldest cores.

Another study has focused on the relationship between job parallelization and temperature in embedded systems [27]. In this context, a job is a collection of tasks that may utilize one or more processors simultaneously. For such a job, a trade-off might exist as to how many processors should be devoted to complete its computational load at a given time instant. Specifically, this study investigates the dynamics for three types of jobs: those which are allocated a fixed number of processors external to the scheduler, those which are allocated a fixed number of processors by the scheduler, and finally, those which can be assigned a varying number of processors by the scheduler throughout their execution. Jobs are sequenced in a periodic task system with deadlines. In such a scenario, if individual threads of a job are spread across multiple processors spatially, the resulting thermal state will be influenced by the relative placement of the threads as well as the operating point chosen for each processor. This interaction is complex, since in dynamically managed multicore systems, individual processors executing parallel portions of the same job may also be at different voltage and frequency settings, resulting in spatial variation in power density. This study, hence, points to the need for an online scheduling algorithm to spread threads of a job among processors and concurrently determine their frequency settings while considering a peak temperature constraint on each processor individually.

Embedded systems present a unique challenge stemming from their operating environment that is generally not observed in other computing systems. Embedded systems can be deployed in environments with vastly variable ambient temperature conditions. Also, for a fixed system these conditions can change over time significantly. A study has focused on this aspect to develop a load balancing scheme under variable ambient temperature [28]. The embedded system is modeled as a set of periodic tasks with deadlines and WCETs and with specifications according to a chip from the automotive domain, where drastic swings in ambient temperature are typical. Such fluctuation in temperature is prone to trigger thermal cycling in the chips largely degrading their reliability. For embedded systems without access to cooling fans and sophisticated dynamic power management mechanisms, thermal runaway also becomes a threatening reality. This particular study proposes a mechanism for the real-time OS (RTOS) to dynamically migrate tasks, such that the real-time deadlines of all tasks are guaranteed, while the system can modify the number of idle cores and the voltage/frequency settings of the cores to minimize the number of thermal cycles experienced by the system. The system utilizes one global ready task queue for the multicore chip. Each core's temperature is sampled from the respective on-chip temperature sensor. In an offline phase, for a number of temperature levels (peak of all cores) a static configuration of frequency settings using minimum number of cores is found such that the task set could be executed without violating their deadlines. This is stored in a configuration table. At run-time, based on the temperature readings, the current temperature level is detected and the corresponding configuration is looked up from the table. Based on the number of cores required at a given configuration, the remaining cores can be placed in idle mode and if the number of active cores in the next stage is less than those in the previous scheduling interval, then the RTOS migrates tasks out of those cores that will be placed in idle mode. The configurations are constructed such that at higher temperatures, a larger number of cores are preferably put to sleep given the timing constraints. Furthermore, at a coarse granularity of time (a multiple of the task graph period) the active cores are re-shuffled in a round robin fashion to further distribute power density uniformly across the chip, so that certain cores are not placed in idle mode more often that others.

While a large fraction of embedded systems execute a well-defined application with deterministic workloads, emerging high-performance embedded systems are becoming versatile in domains with workloads that reveal their make-up only at run-time. A study has focused on the thermal issues in a multicore embedded system with such non-deterministic workloads [29]. Concepts from real-time calculus [30] are utilized to derive a new estimation for the WCET of each task based on the time varying workloads. The multicore chip is assumed to contain a homogeneous set of PEs. The thermal considerations are incorporated to the task assignment step. PEs are assigned with a clock frequency to perform the total estimated workload within a deadline prior to task assignment. Next, for each task an assignment problem is formulated in the form of a non-linear binary integer program. The main consideration is that for each task a destination PE must exist such that the tasks estimated bound on the execution time can be met. While performing this assignment, the thermal consequence of an assignment on the peak temperature across the PEs is estimated with an equivalent thermal RC-network model of the chip. It is noted that, evaluating the thermal state per each potential assignment being evaluated would incur an excessive computational overhead not suitable in real-time applications. A metaheuristic is proposed to perform the thermal evaluation efficiently by combining multiple related assignments in one evaluation step. This is achieved in two phases. First, the temperature estimation of each PE is decomposed into two terms: an assignment-independent component and an assignment-dependent component. The assignment-independent component is calculated a priori once for all assignments. As for the assignment-specific temperature term, it can also be simplified by observing the difference between two assignments. Several assignments will only differ by a few tasks being placed in different PE, while the remainder might completely overlap. In these cases, once the temperature is estimated for one assignment, the next similar assignment's temperature is calculated incrementally by only updating the non-overlapping tasks.

The task assignment algorithm described above might incur inaccuracies in the temperature estimations, especially due to the simplifications introduced to manage the computational overheads. These might be a necessity for embedded systems without on-chip sensors. However, many modern multicore chips deployed in embedded systems are also equipped with sensors. These sensors can be leveraged to enhance and refine the temperature estimation routine used in this method.

Another study similarly challenges the traditional assumption of an embedded system being tailored for a single application and proposes a scheduling method to manage tasks from a mix of diverse applications [31]. For instance, mobile communication devices such as smartphones multitask, executing applications such as audio, video, web browsing, etc. Different application characteristics result in variability in the thermal response. Performance goals such as throughput and fairness among different applications for their access to the computation resources are pursued while maintaining the operating temperature of the system under a threshold. Another differentiating property of this system is that some tasks in the mix have soft real-time deadlines in a periodic fashion (e.g., video processing tasks with a frames per second repetition) while others have even higher tolerance to delays and drops in quality (e.g., word processing) and they are classified as best effort tasks in this study. The basic approach uses a mechanism to estimate the thermal impact of a task and prioritize those tasks, which will incur a smaller temperature increase in the system. At the same time this mechanism pursues the trade-off between throughput and fairness. A predictive temperature model is used, which captures the expected temperature rise of a system if a task executed for a period of time t into the future, if the current temperature (i.e., initial temperature T_{init}) is known. The temperature reached within t time of execution is modeled as shown in (5.10).

$$T(t) = T_{ss} - (T_{ss} - T_{init}) \times e^{-Kc} \tag{5.10}$$

T_{ss} is the steady-state temperature that would have been reached if this same task executed for an indefinite period of time. T_{ss} can be projected by observing the temperature T_c after a specific amount of time c after starting execution from a known T_{init} using on-chip sensors. This relationship is depicted in (5.11):

$$T_{ss} = \frac{T_c - T_{init} \times e^{-Kc}}{(1 - e^{-Kc})} \tag{5.11}$$

κ is the hardware-specific thermal constant of the underlying processor system. This model is used by the online scheduler to evaluate tasks with soft deadlines and predict whether the execution of a selection of tasks would be permissible at a given instant without exceeding the thermal threshold of the system.

The basic scheduler is composed of two specific scheduler components for soft real-time tasks and best effort tasks. For each task an execution time frame is predicted by the scheduler and used to determine the sequencing of tasks. Given a schedule for the soft real-time tasks as such, a refinement phase evaluates the thermal effects. At a given time instant, there are tasks currently executing on the embedded system (current tasks) and other tasks that are under consideration for scheduling (future tasks). At each scheduling interval the scheduler aims to optimize the following. Given the predicted execution time of the future tasks and their corresponding predicted steady-state temperatures, it determines the maximum initial temperature T_{req}, at which they can start their execution and still not exceed the temperature threshold upon completion. In order to meet this constraint, the voltage and frequency of the current tasks can also be adjusted to help bring the initial temperature for the future tasks to the desired level. Similarly, the voltage and frequency settings of the future tasks can be adjusted such that starting from the initial temperature upon completion of the current tasks, they do not exceed the maximum allowed limit once they also complete execution. At the same time, the soft real-time scheduler identifies any thermal headroom available in the system. If the temperature at the completion of the future tasks that are being scheduled is below the maximum limit, the corresponding slack is relayed to the second scheduling component in charge of the best effort tasks. Utilizing this information, the best effort scheduler can adjust the priorities among different tasks. Specifically, it classifies best effort tasks into two types: cold tasks and hot tasks. Cold tasks are those that would not exceed the target T_{req} set for the next set of tasks being currently scheduled. Hot tasks are those that cannot meet the T_{req} goal. The best effort scheduler aims to split the execution time slice allocations between hot and cold tasks in an alternating fashion. Each task is assigned a CPU share initially. The run queue is then split into the hot and cold task sets. The online scheduler selects a task from the cold task set with the smallest CPU share and similarly one from the hot task set with the smallest CPU share. Depending on the predicted thermal impact of each individual task, the CPU share is adjusted with a parameter ranging from 0 and 1. For the best effort tasks it is desirable to execute them at the non-scaled voltage and frequency settings. If the share received by a hot task under this setting is below a minimum share limit (determined by the scheduling system) only then the run-time system considers additional fairness measures. The share of this task is increased with a corresponding scaling of its voltage and frequency setting, so that the task can still be scheduled albeit at a lower performance and maintain the thermal constraints as well.

Finally, another classification that can be associated with embedded systems is distributed computing features. A study modeled a distributed embedded system comprised of a wide variety of heterogeneous PEs (e.g., CPUs, FPGAs, DSPs) [32]. A thermal model similar to the system described above [31] has been used in this study. One new aspect of this investigation is that each PE in the distributed embedded system samples its own on-chip temperature sensor and broadcasts it to all other PEs through the common communication bus periodically. Each PE maintains a sorted list of temperatures of all PEs in the distributed system. When a

PE experiences a thermal emergency, it initiates a negotiation with the coldest PE in its list via a handshake signal and requests to swap its workload with the colder PE. In the subsequent time period data transfer occurs to exchange data between PEs and each PE resumes executing the other's schedule. During the data transfer, a PE may miss its deadline for a real-time task that was assigned to. This is mitigated by allowing the task with the earliest start time (or deadline) to use the communication medium for its data transfer. Overall, this method provides a practical approach to co-managing thermal constraints and migration overheads to still maintain the real-time deadlines of tasks executing on this embedded system.

5.3 Thermal-aware virtualization

Virtualization is a powerful tool to allow cost-effective and seamless sharing of a large pool of physical computational resources by a diverse set of OSs in a large-scale system [33, 34]. Drawing its roots from implementations for specialized servers, it has become a widely used method with commercial success from server systems in datacenters and other high-performance computing platforms. Virtualization also finds use in desktop computing for dual boot systems, typically for co-hosting Windows and Linux OS.

While a single system's physical resources (processors, memory, I/O, networking) are managed by its OS, in a virtualized system an additional layer is deployed to oversee the management of the physical resources. This layer is referred to as the virtual machine monitor (VMM). The VMM coordinates the access of local OSs to the physical resources, each provided with the specific interfacing needs.

VMMs present a suitable medium to facilitate methods such as migration and load balancing that are highly relevant to creating uniformity in activity, power, and energy consumption at a system-wide scale. Recognition of this opportunity combined with the pressing need of optimizing the energy and thermal profiles of datacenters and other large-scale high-performance computing clusters led to increasing interest in incorporating thermal awareness into virtualization tools.

A comparative study evaluated the benefits of virtual machine migration against voltage and frequency scaling for thermal management in high-performance computing systems [35]. The system-level thermal management methodology envisioned as four layers of interacting abstractions as illustrated in Figure 5.6.

The topmost layer of abstraction represents the user(s) of the datacenter and their applications and workloads. The main performance metric at this level would be QoS. The second layer denotes virtualization as the overarching mechanism for accessing the datacenter. Events such as VMM migration and load balancing occur at this level and the success would be measured in terms of performance and energy consumption of the datacenter and the subsequent operating cost. The third layer corresponds to the physical resources of the computing system. For power and thermal management, techniques such as dynamic voltage and frequency scaling, power capping to processors and/or cores, and clock gating are available at this

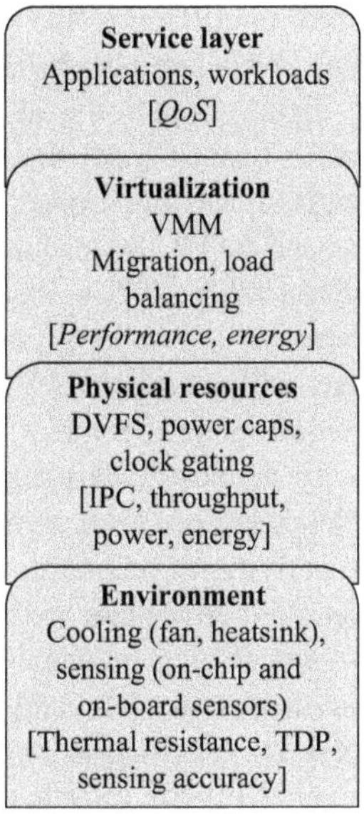

Figure 5.6 A system view of interacting layers of abstractions involved in power and thermal management in a datacenter

layer. The bottommost layer represents the physical environment of the datacenter, including the cooling system, temperature sensors that help track the thermal state within chips, racks, cabinets, and rooms, placement of the cabinets and corresponding cooling distribution in the room, dimensions, etc.

Several methods are implemented at the virtualization layer for thermal management in this study [35]. First, the system responds to temperature emergencies reported by temperature sensors via migration. The main objective is to drastically minimize the CPU activity at the location of the hotspot. This is achieved by migrating the VM hosted on the server where the hotspot was detected along with the host OS and all applications running on this resource. This migration can occur at varying amounts of overhead depending on various factors. For instance, depending on whether an active server is readily available to receive the migrated components or a server in sleep mode needs to be booted up, the overhead of the thermal intervention and its impact on performance will be different. Also, the extent the applications being migrated will create resource contention at the destination server will determine the performance degradation at the destination. In the extreme case of no server being able to sustain the additional thermal stress (they may be already operating near the thermal emergency threshold, or the impact of migration can be predicted) induced by migration, the hot server may simply be suspended to mitigate the heat accumulation and wait until a destination server becomes available.

Another aspect of VMM management that can be leveraged for thermal management is the way virtual CPUs are assigned to actual physical CPUs. The VMM can determine this mapping based on utilization and performance concerns. On the other hand, certain select virtual CPUs can also be explicitly assigned to specific physical resources by setting affinity relations, also called pinning, if special workload properties are known to benefit from this action. Often this is used to exploit data locality. In this study, it has been proposed to leverage pinning as a means to protect thermally stressed CPUs. Specifically, VMs are pinned to other CPUs, preventing them to migrate to those CPUs under protection, so that the local thermal management mechanisms of these CPUs can lower their frequencies and alleviate their thermal emergency without a hosted VM forcing the CPU to run at a high frequency to meet a QoS requirement. On the other hand, it must be noted that this will then intensify the activity on those CPUs subject to pinning. Other studies pointed out that system-level interventions to activity management should avoid the pitfall of causing excessive activity on one core, which will in turn prevent all cores on that CPU to run under their true peak frequency, since most CPUs still deploy chip-wide clock and voltage scaling mechanisms [36].

Thermal management solutions at the OS- and VMM-levels present unique opportunities due to their access to information concerning the workload, QoS, deadlines, priorities, etc. However, there are interface barriers to be overcome and response time can be a limiting factor. A study has probed into the boundary in search of a thin virtual layer that is not necessarily part of the OS, but rather an entity that is intimately connected to a VMM as well as the underlying hardware [37]. This virtual layer is referred to as the virtual thermal manager (VTM). A VMM is used to pass workload information and invoke and suspend the VTM periodically (every 10 ms), yet the VTM is not visible to the OS. VTM collects temperature sensor data and maintains temperature history tables of active threads on the system. It also maintains the thread to core assignment information from the VMM. A detailed correlation data between thread properties and thermal behavior is maintained by storing the most recent two readings of all on-chip temperature sensors for each active thread. VTM evaluates thermal gradients between cores and peak temperatures of cores. Upon detection of a gradient larger than a threshold between two cores, VTM initiates thread migration. This VTM mechanism has also been extended with capabilities to perform architectural reconfigurations, for example, issue throttling, reduction of the issue width, and retire width.

Thermal management methods for multicore systems usually pursue various types of activity movement in order to better distribute heat dissipation across the total landscape, be it one chip or a number of servers. While the guest OSs apply their own policies for migrating activity, in the case of virtualization, the VMM would also invoke additional measures to manage thermal balance at the system level. Potential conflicts among the policies of multiple virtual CPUs and the VMM and the difficulties of a virtual CPU thermally managing its currently assigned physical resource have been subject to further investigation [38, 39]. There are two main approaches to balance power consumption and prevent hotspots occurring early and often. The first solution, which is equally applicable to non-virtualized systems, is to migrate tasks on a virtual CPU from one core to another within the same resource. This particular study proposes a method to cater to the second option. Threads running on one virtual CPU are exchanged with another virtual CPU entirely. This requires a fine grained prediction of the power consumption for each virtual CPU. The prediction needs to be frequent enough to capture the highly time varying aspect of a virtual CPUs power, since the CPU can execute multiple guest OSs over multiple time slices with diverse behavior. The issue is that the VMM has knowledge of the currently executing tasks and the history to some extent, yet, the next task to be executed is determined by each guest OS respectively. Hence, in this particular study, a migration policy update with its corresponding power and temperature predictions are suggested to take place at a rate faster than the thread scheduling ticks of the guest OSs. A temperature and power estimation method using performance counters has been employed, arguing that thermal diode-based sensors could not be sampled at the required frequency to update current temperature. This limitation can be easily overcome with the latest generation on-chip temperature sensors, which deliver high sampling rates.

Given a mechanism, which can estimate the power consumption and resulting temperature state of a virtual CPU, in broad terms the problem of thermal balancing boils down to identifying the virtual CPU that is consuming highest power and swapping its physical resource with the virtual CPU consuming lowest power. There are two main types of migration defined in this framework: (i) intra-node migration and (ii) inter-node migration. Intra-node migration does not involve any displacement of the VMM. Only the mapping of the guest OS to the physical resource is changed dynamically by the VMM, transparent to the guest OS. When performing inter-node migration, the VMM is also relocated to the destination node. Various optimizations have been applied to this basic framework to perform frequent migration in a cost-effective manner. One of these optimizations is the batch migration to deal with cases where a virtual CPU may be executing a large number of threads from a high number of guest applications executing on it. This indicates that, not each virtual CPU migration will involve the same number of threads to be migrated. In the case of a large thread collection, the environment has been extended with a user-level scheduler component that is capable of migrating a number of threads in a single attempt.

The concept of hardware virtualization is also applicable in other types of systems. One example is PE virtualization in networks on chips (NoCs) [40]. A NoC is comprised of a collection of PEs interconnected with routers as depicted in Figure 5.7. Each physical PE can be virtualized to host more than one processing task and coordinate among them via context switches.

One interesting aspect of virtualization in the context of NoCs is the variability in communication cost depending on the mapping to physical PEs. Communication delay is not only a function of interconnect, which is relatively uniform, but also the number of routers that a packet needs to travel through, which is highly variable. This variability in the routing paths also impacts energy consumed for communication, which is a significant fraction for NoCs. A method based on the genetic algorithms paradigm has been introduced to enhance the thermal awareness of the PE virtualization and task placement considering these effects [40]. The virtualization of physical PEs and placement of logical computation tasks as

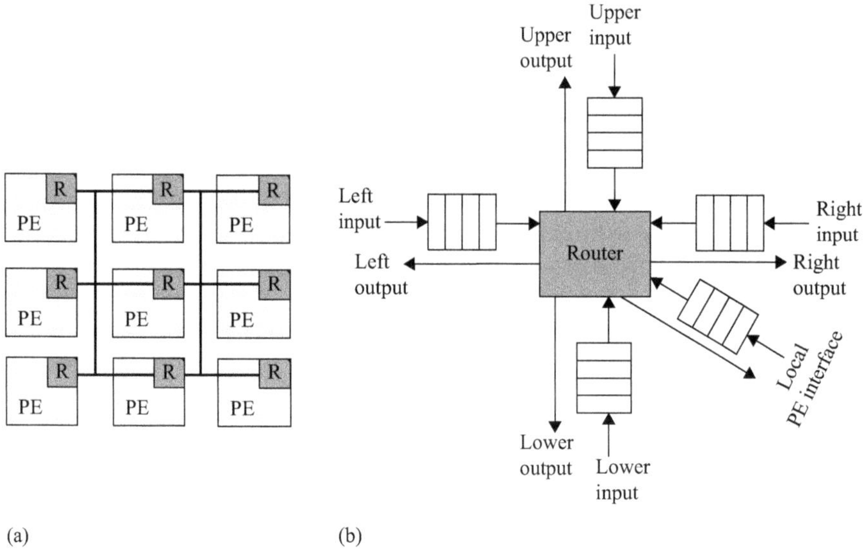

Figure 5.7 Abstract illustration of a network on chip architecture: (a) most NoCs are implemented as a regular array of PE tiles with a local router to manage inter-PE communication; (b) a router implements a specific algorithm to forward incoming data packages to a particular output port, inputs can be buffered as shown in this figure

virtual PEs is represented with an initial random set of binary encoded solutions. This solution set comprises the initial population of the genetic algorithm. Each solution is evaluated with the fitness function, which helps to assess the chance of survival for each chromosome encoding, that is, solution. In this case, the fitness function represents the uniformity in temperature distribution, which correlates with the peak temperature experienced by the hotspot. This is quantified as follows. Each PE location with a given task assignments per the encoding is associated with a representative temperature. Within a $t \times t$ region of the PE tiles the sum of temperatures of all PEs are computed. The fitness function is then defined as:

$$Fitness = \frac{1}{MAX(S_{t\times t}|\forall\ t \times t\ regions)} \tag{5.12}$$

The inverse of the maximum temperature sum across all $t \times t$ regions defines the fitness of a given chromosome. The $t \times t$ region signifies the range of strong thermal coupling in the lateral dimension across the NoC. The genetic algorithm evolves the population of initial solutions by two mechanisms: crossover and mutation. Using crossover, off-spring is generated from two chromosomes. Each chromosome is associated with a probability for crossover. Mutation is realized with random swapping or shifting, which also occur with pre-defined probabilities. Tuning of these parameters is crucial for the success of the genetic algorithm. The population size is another parameter, which impacts the final quality and total execution time of the algorithm in opposing ways. Finally, a termination criterion needs to be defined, which can be the total number of iterations or the minimum amount of incremental improvement expected in the fitness function.

5.4 The role of the application layer in shaping thermal profiles

Applications drive computing systems and their nature clearly shapes the thermal state of the underlying system. Users and the applications that they demand define the performance expectations and QoS levels. These metrics define the power density in computing systems and also they influence the economic dynamics of the user-service provider-manufacturer ecosystem.

Thermal awareness has already penetrated into the topmost layer of abstraction in the system, including compilers, workload management, equipment characterization in datacenters, etc. Various code optimization and analysis methods from a thermal point of view have been proposed. An early study evaluated existing source code transformations to quantify their thermal impact [41]. Based on the observations made in this evaluation some guidelines towards a thermal-aware compilation flow are laid out. A microcontroller is instrumented to collect power measurements. The block-level floorplan of the microcontroller is converted to an equivalent thermal circuit to translate power measurements into temperature, at the same time, block-level utilization statistics are used to derive an analytical relationship between power and functional unit utilization. An image filtering benchmark has been used as a case study, which is representative in terms of the activity and thermal stress experienced by an embedded system. Three code transformations have been compared in this study, which are loop splitting, dynamic memory, and loop pipelining. Loop splitting breaks down a loop into multiple loops that iterate over different subranges of the original loop such that some dependencies in the main loop can be eliminated to improve performance. Dynamic memory allocation can be performed with compiler directives to optimize data locality at run-time. Loop pipelining involves re-ordering of instructions in a loop body and managing dependencies such that a maximal number of iterations can be active simultaneously.

The impact of these compiler-directed optimizations have been evaluated from two perspectives: (i) the overall peak temperature of the system and (ii) localized gradients among different processor blocks. It has been observed that optimizations improving the execution time generally have a positive impact on peak temperature, since the activity spans a shorter duration of time and does not allow for significant rise in temperature. Although loop pipelining reduces power consumption compared to the baseline case, the peak temperature is higher. This supports the arguments of several other past studies, which point to the need for thermal-aware optimizations beyond power minimizations, since lowest total power does not always correspond to lowest power density [42, 43]. Investigation of gradients reveals more insight into the impact of the compiler and application behavior. This study evaluated two gradient-related metrics [41]: largest temperature differential among all pairs of processor blocks and typical gradient between all pairs of blocks. The largest differential can be helpful in deducing the expected thermal cycles and stress experienced by the chip. In this study, loop pipelining reveals the largest gradient, which occurs between the register file and the accumulator register, where intermediate arithmetic results are stored. During the execution of a pipelined loop, drastic transitions across the memory space exercises special registers in the register file causing distinct hotspots and the large swing in temperature with respect to the rest of the chip.

One of the early proposals for thermal-aware compiler methods focuses on fitting computation into very large instruction word functional units [44]. Functional units are known to exhibit chip hotspots often. This method introduces compiler-directed scheduling of tasks among functional units in a balanced way such that power density distribution across them will also be uniform. The temperature of functional units needs to be approximated at this early stage of the flow. In this study temperature of a thermally decoupled block is approximated using the estimated power consumption per instruction P, and three physical parameters of the underlying hardware: thickness of the die t, thermal conductivity of silicon κ, and area of the functional unit A:

$$T_{FU} = P\frac{t}{\kappa A} \tag{5.13}$$

Furthermore, heat diffusion between two adjacent functional units is modeled as:

$$H(FU_1, FU_2) = s(T_1 - T_2) \tag{5.14}$$

Assuming that all functional units occupy same area and the material properties are same across the chip, using the relations shown in (5.13) and (5.14), it has been noted that the heat diffusion between adjacent units is proportional to the difference between their power consumption. Hence, the optimization objective is formulated as balancing the power density across functional units. Since functional unit have same area, this was simplified further to dynamic power per unit. This is achieved by developing a compiler pass to modify an initial performance-driven schedule of arithmetic instructions. The proposed compiler pass first evaluates the total switching power for each functional unit under the given high-performance schedule. It also calculates the average switching power per unit under perfect balancing using the total power of all units divided by the number of units. The main idea is to strive to achieve this ideal power consumption for each unit. The compiler considers units in sorted order of their power consumption and looks for opportunities to reschedule computation from the unit with highest power consumption to the unit with the lowest. A rescheduling move is allowed only if the dependency constraints are still met. This method is also extended with consideration of thermal-induced leakage using a technique called IPC tuning. This refers to program phases where IPC is lower than the number of functional units on the system. In response to this, a fraction of functional units can be power gated to reduce leakage power. This basic technique is customized in this compiler scheme to avoid keeping the same few functional units active across a large number of iterations of a basic block or across multiple loops with IPC tuning opportunity. This is achieved by rotating the power gated and active functional units at the loop-iteration granularity or as the code transitions from one loop to the next.

A code analysis framework has been proposed for parallel code targeting high-performance computing systems to identify potential thermal bottlenecks [45]. This software tool leverages on-chip temperature sensors to correlate real-time measurements with the software program. Such a tool can be effective in pinpointing specific locations in a parallel code, which could best benefit from thermal-aware optimizations. Many high-performance computing applications are carefully hand-tuned by the developer for performance, hence, it is acceptable to have significant programmer input in the design process. Within the same framework, developers can be guided towards (or away) from certain optimization choices for thermal management. Also, such instrumentation can facilitate testing the behavior of a given parallel code across different machines in a cluster in a large-scale system. In broad terms, this profiling tool monitors a program at the function level and collects several temperature readings from sensors throughout the execution of the function. The profiling tool then maps the function timestamps to temperature readings.

A review on the needs for thermal management and future trends, various compilation techniques, such as resource hibernation and spatial scheduling, have been suggested for power, energy, and thermal management [46]. Compiler-directed scheduling of instructions can help to favor or skew the duty cycles of certain resources to avoid their activities to overlap or allow a sufficiently long stretch of idle time to remove accumulated heat in sleep mode. Spatial scheduling, similarly, can take the thermal coupling between physical resources into account and spread the scheduling of instructions such that those that utilize thermally coupled resources have minimal overlap in their execution intervals.

Two prominent compiler optimizations: register allocation and loop unrolling have been revisited with thermal considerations in the foreground [47]. The main idea is to spread access activity across multiple registers in a uniform manner by leveraging loop unrolling. A high level thermal cost model has been derived to correlate a register use to the expected increase in temperature of that register and the physically adjacent ones through conduction. For a given register allocation, the predicted steady-state temperature of each register is evaluated. Then, the registers are categorized into two types; hot and cold. Those with a temperature prediction above a certain limit are classified as hot registers. Loop unrolling is then used to re-distribute the activity of these registers. For each inner loop, a set of alternative registers are sought that are currently not utilized by this loop. Starting from the "hottest" register, variables are then moved over to these alternate locations. Register allocation is performed in this fashion each time a new unrolling has been applied to a loop.

A detailed evaluation of high-performance computing workloads aims to establish specific guidelines according to the behavior of each application and tailor power and thermal management from an application-centric perspective [48]. Profiling of the workloads with a breakdown of the behavior and phases of different sub-systems helps to better assess the significance of a given sub-system for a given workload. Thereby, for a given workload the usefulness of a power-saving strategy (e.g., low power modes, P-States) can be estimated and it can be judiciously applied only for those sub-systems that are consequential in terms of power and energy. For I/O-intensive applications CPU utilization is much lower than the rest of the target applications, where P-State choices at the lower frequency ranges are almost always possible. For some other workloads the network interface card has certain special bursts of activity with idle periods elsewhere. The intensity of synchronization in a workload affects its CPU, I/O, and memory utilization. Based on these evaluations and a cost metric of switching each sub-system on and off (or step up/down in frequency), the relative gain in energy is evaluated with a high level model. This resulting model can also find uses at run-time in a predictive and dynamic power management scheme.

5.5 Thermal-aware optimizations in datacenters and supercomputers

Cooling is the second largest expense of well-designed datacenters, with costs ranging between 30% and 50% of the total electricity bill. The cooling systems in a typical datacenter are similar in the basic functionality to those used at the chip and board-level. Forced air cooling as well as water-liquid cooling is heavily utilized in individual floors or rooms. Water cooled systems either deploy a chiller plant to actively pump water at a lower than ambient temperature from a central cooling towers or use warmer air in chiller-less systems. In either case, coolant water is piped to the computer room air conditioning (CRAC), also referred to as the heating ventilation air conditioner, units to interact with the air circulation. Air supplied to the server cabinets then helps to absorb the heat radiating from the heatsinks of the server chips. There is an industry focusing on delivering building and room cooling solutions for datacenters. At the same time, there is a large body of work focusing on mitigating the causes of heat dissipation as well as enabling more effective and cheaper heat removal with new system and application-level datacenter management.

5.5.1 Characterization of heat dissipation in data centers and performance metrics

Some studies have focused on understanding the thermal interactions within a datacenter by monitoring the airflow as it carries heat away from the servers. Use of distributed sensors has been proposed to this end [49]. A typical datacenter arrangement which results in the following flow of hot air has been considered in this study: the cooling airflow supplied by the CRAC units enters through the front sides of racks and heats up as it travels though the chassis hosting the servers and interface chips. The hot air is extracted to the return loop by CRAC intakes. Similar to the conductive thermal coupling within a chip between physically adjacent blocks, at the datacenter level there are also coupling interactions. Hot air exiting one aisle can escape through the inlets of racks in another aisle. By careful monitoring of the on-board inlet and outlet sensors, this complex interaction can be demystified and preventive measures can be implemented. A thermal model has been constructed to capture this phenomenon. Using a given power distribution across racks at a time, this model can then be used to predict locations where cooling performance will be less efficient and server locations, where more aggressive thermal mitigation should be adopted locally to maintain the uniformity in outlet temperatures of all servers. This will help to avoid penalizing all servers or boosting up cooling globally for the entire datacenter. A one-time offline CFD-based simulation model is created to characterize re-circulation using data collected from on-board sensors of the servers. By fitting these data with the CDF modeling tool, cross interference coefficients are derived for a statically designed datacenter, for a representative workload. These coefficients are then used in compact analytical models that can be applied at run-time for fast prediction of outlier servers with high inlet temperatures [50].

Another similar study used offline experiments to characterize the thermodynamic interactions inside a datacenter and also proposed a specific method to schedule workloads by taking the temperature distribution into account [50]. This temperature-aware workload scheduling algorithm pursues two main directives: (i) ensuring that all server inlet temperatures are constrained by a safe limit and (ii) the required temperature of the cooling air is maximized, such that the CRAC efficiency is enhanced and its operating cost is minimized. Similar to the aforementioned method, the key to effective management of inlet temperatures is minimizing heat re-circulation from hot servers into colder ones. Instrumenting the datacenter with thermal sensors, the amount of heat generated by server activity, and the amount of heat removed is measured. Furthermore, from these quantities, the amount of heat re-circulating in the datacenter is also derived for a reference baseline workload. Several servers are clustered into a so-called pod to limit the complexity of the calibration run and be able to generate a measurable amount of heat locally to distinguish the impact of re-circulation. For each pod j the heat re-circulation factor (HRF) is calculated as follows:

$$HRF_j = \frac{Q_j - Q_{ref}}{Q_j^{rec} - Q_{ref}^{rec}} \tag{5.15}$$

Q_{ref} and Q_j are the heat generated by a reference workload in a baseline experiment and heat generated by a second experiment maximizing CPU utilization of the servers in that particular pod, respectively. Q_j^{rec} and Q_{ref}^{rec} denote the amount of heat being re-circulated, during the reference experiment and the second experiment, respectively. If no heat was re-circulated, all inlet temperatures would be equal to the air temperature supplied by the CRAC. The non-zero difference measured by temperature sensors in practice, however, helps to derive the amount of re-circulated heat.

The proposed workload scheduler distributes the power budgets to each pod in proportion to the fraction of its HRF out of the total HRF from all pods. Hence, servers that are at a disadvantage in terms of cooling due to topological reasons or due to highly active neighboring servers are less loaded to reduce their local heat dissipation. This study has later been extended to include on online learning-based technique to eliminate the offline calibration step [51].

Air cooled systems such as those discussed earlier use cabinets for servers are typically constructed with raised floors, open inlets at the bottom and outlets at the top, hence, the issue of re-circulation. Water cooled systems, on the other hand, are constructed as sealed enclosures accessible with water pipes connected to a chiller unit and pumps. A systematic thermal characterization of such a water cooled system was presented using temperature sensors reporting CPU temperatures to evaluate the efficiency of different coolant flow rates [52]. A server cabinet was populated with 16 1U Dell servers dissipating a maximum power of 450 W. Cabinet mounted fans circulate cooling air within the cabinet. The hot air returned by the heatsinks of the servers is routed through a duct to an internal heat exchanger. The efficiency of the heat exchanger has been quantified as the ratio of two temperature differentials: (i) temperature difference between the air outlet and inlet; (ii) temperature difference between the water inlet and air inlet:

$$\varepsilon = \frac{T_{out}^{air} - T_{in}^{air}}{T_{in}^{water} - T_{in}^{air}} \tag{5.16}$$

This study reveals that as the water flow rate is decreased to 57% and 42% of the maximum, the heat exchanger's efficiency dropped from the baseline 0.83 to 0.75 and 0.66, respectively.

Design and modeling of the cooling main infrastructure, such as the heat exchanger studied in the study described above are essential for cost-effective thermal planning. Another study provides a broader system-level view on the design guidelines and optimization objectives for temperature-aware data centers [53]. The first observation is that each data center has an organization-specific thermal blueprint, much like a processor chip having its commonly occurring locations of hotspots. If these locations are identified, ordering of cabinets and classification of racks into alternating hot and cold aisles can be performed to de-skew this imbalance with proper workload assignment. Overheating and excessive stress on the cooling unit can also cause cooling failure and expensive downtime. An intelligent data center management system could relocate workload from racks that are likely to experience downtime due to such failures. This work argues that in order to enable any of these aforementioned techniques, accurate monitoring with precise correlation of thermal events to system events is a necessity. This study also points to the important role that VMMs will play to facilitate thermal-aware workload management.

In fact, a large number of investigations focused on workload placement to control heat distribution within data centers. Similar to the characterization efforts discussed earlier, most of these methods rely on characterization of the data center with thermal monitoring. One approach uses monitoring to designate certain locations of the data center as being easier to cool, such that more intensive workloads could be concentrated in those areas [54]. This workload placement method is motivated by anecdotal data from within a commercial data center and experimentation on academic batch job servers, which strongly indicate that there is a significant fraction of low utilization periods. This supports the presence of opportunities to move workload around in the data center without resource contention concerns. A new efficiency metric is introduced to guide the workload placement. This metric helps to define the cooling dynamics stemming from the topology of the data center (outlets, inlets, vent tiles, separation between aisles, etc.) and the CRAC placement in the data center. The thermal correlation index is then defined as follows [54]:

$$TCI_{i,\,crac\_j} = \frac{\Delta T_i}{\Delta T_{crac\_j}} \tag{5.17}$$

This index indicates the temperature change measured at the *i*th rack inlet in response to a ΔT_{crac_j} temperature change induced by the *j*th CRAC. This metric can be evaluated for each CRAC in the data center using temperature sensors with a calibration run. Using this metric and the concept of hot air circulation assessed by several similar studies, a more general index, named the local workload placement index (LWPI), is derived to determine the benefits of a certain workload placement for each server *i*:

$$LWPI_i = \frac{(T_{set} - T_{in})_i + \sum_j \left[\left(T_{airSupply} - T_{airSupply},\, min \right) \times TCI_i \right]_j}{T_{in,\,i} - T^{eff}_{airSupply,\,i}} \tag{5.18}$$

This metric utilizes measurements collected from a number of built-in temperature sensors located near each server *i* and CRAC units to compute the following terms:

$T_{in,i} - T^{eff}_{airSupply,\,i}$: the difference between the current server inlet temperature and the temperature of the air delivered to this location from the closest vent tiles.

$T_{set} - T_{in}$: the difference between the pre-set maximum temperature limit for the sensor inlet temperatures and the current inlet temperature measured by the server's sensor.

$T_{airSupply} - T_{airSupply}, min$: the difference between the air temperature supplied by a CRAC *j* and the minimum achievable air temperature for the CRAC units. This difference is computed for each CRAC unit individually and a weighted sum is computed using the respective TCI defined between server *i* and CRAC *j*.

This metric has been used in determining the placement for CPU intensive batch jobs specifically. A static schedule has been derived using job length estimates. Among the compute intensive jobs, those have the longest running schedules have been then assigned to servers with the best LWPI metric.

Another method focused on optimizing inlet temperatures for energy optimal task assignment using a linear heat re-circulation model [55]. The task placement is then solved by two alternative methods using a genetic algorithm approach and a sequential quadratic programming approach, respectively. This method derives a coefficient of performance metric for a chilled water cooled data center as a function of the air supply temperature:

$$COP(T) = 0.0068T^2 + 0.0008T + 0.458 \tag{5.19}$$

Since the energy spent by the CRAC unit to provide cooling is proportional to the air supply temperature, running a system at the minimum air supply temperature while no rack's inlet temperature exceeds a certain threshold is expected to provide the most energy efficient cooling solution. The point at which a server rack's inlet temperature exceeds a limit is also referred to as redlining. Workload placement within a data center is modeled as placement of a task that utilizes c_i blades in a server chassis that contains m ($m \geq c_i$) blades in total. The power dissipation within a chassis i due to this placement is modeled with a linear function, using a baseline idle power consumption term b for each chassis and a constant power consumption term a per active blade:

$$P_i = b + c_i a \tag{5.20}$$

For a given task placement vector $\boldsymbol{c}$, the inlet temperature vector of all chassis in the system is then modeled with a linear combination of the respective nearest CRAC supply temperatures and the power consumption vector of the data center. This model thereby takes both the local heat dissipation within a chassis i and the secondary influence of re-circulated hot air from all other chassis into account as follows:

$$\boldsymbol{T}_{in}(\boldsymbol{c}) = \boldsymbol{T}_{AirSupply} + \boldsymbol{Db} + \boldsymbol{Dca} \tag{5.21}$$

A dynamic learning method has also been used to infer the cyberphysical structure of the data center cooling system to make predictions on server temperatures [56]. This method helps to alert the management system on imminent thermal emergencies well in advance, so that preventive measures can be engaged sooner.

5.5.2 Software environments and provisioning at the system-level for thermal-aware management

Given our understanding of the physical characteristics of thermal behavior in a data center and the available software tools to manage and re-shape workloads, several proposals focused on efficient implementations of these management environments. One proposal was to create a seamless merger between the physical infrastructure and the software system managing the data center [57]. Two major functionalities at the system level, namely, the task scheduler and the process migration tool have been enhanced to consider the self-heating within servers and heat re-circulation among servers into account. The environment leverages on-board inlet and outlet temperature sensors and builds a sensor readings database. The historical data collected in the database is used for short-term decision making by the scheduler as well as by an external policy controller to assess trends at larger time scales. The scheduler can implement new policies leveraging inlet temperature data, such as gradually decreasing the priority of a server to receive an incoming task as its inlet temperature increases. Job pre-emption from a node can also be tied conditionally to a node's inlet temperature exceeding a certain threshold. At the high level, trigger limits for various thermal-aware actions can be continuously monitored and adapted by the policy controller at longer decision intervals.

The pinning method for influencing task affinity towards select servers has been explored as an option for activity balancing for reactive thermal management [58]. In essence, servers that require an intervention due to excessive temperatures reported by their inlet temperature sensors are prevented from additional computation load for a certain period of time by pinning compute intensive tasks to other servers.

The task of determining the active server set at a given time for achieving the best thermal profile leading to minimal cooling energy has been formulated and further studied with a systematic approach [59]. This problem is also referred to as server provisioning. This technique is also useful when aiming purely for power savings in a data center. Provisioning a sufficient amount of servers to meet a QoS demand, while the remaining servers can be put to idle mode for power savings is an effective method. Incorporating thermal awareness into this scheme makes it serve a new yet related purpose. In this particular study, the problem of thermal-aware provisioning problem has been formulated as a nonlinear min–max binary integer programming and attacked with various heuristic approaches. While the major decision variable is a binary variable to denote whether a given server is selected or not for a given epoch, the non-linearity is introduced to the problem due to the nonlinear nature of the energy efficiency of the CRAC system as a function of server activity.

Data centers are at the center of attention in the high-performance computing systems domain due to their commercial significance and wide spread deployment. Technological advances in large-scale computing are pushing towards exascale systems. Modern supercomputers are predecessors of these systems and they are evolving at a rapid pace with a high number of competing systems around the world supporting significant scientific discoveries, drawing societal attention. Supercomputers naturally suffer from the same cooling energy overheads and reliability concerns as the commercial data centers. Perhaps, thermal management has become more visible for data center-related studies, since the cost of ownership for enterprises is a more pressing concern. Nevertheless, the reliability and performance degradations caused by thermal management in modern multicore chips deployed in supercomputers are starting to threaten the peak performance expected from these systems. At the same time, a movement towards higher energy efficiency in supercomputers is inevitable, as it is evident from many recent systems considering chiller-less water cooling systems [60, 61].

A recent study introduced an analytical thermal model for a hot water cooled supercomputer and a control system to manage the cooling system with minimized pump power [62]. The controller manages the optimal temperature for the coolant water to exploit opportunities to operate the cooling system with warm water instead of chilled water for as long as it is feasible. The system dynamics are expressed with an analytical model in a holistic fashion by summarizing the main activities at two nodes: (1) the heat exchanger between the computing system and the pump; (2) the heat exchanger on the return path at the evaporator. These two structures are represented with effective thermal resistances conducting the heat movement in a system of equations that ultimately describe the temperature differentials between the inlet and outlet coolant temperatures. The control parameters, such as the flow rate, valve position, and the minimum coolant temperature supplied by the chiller, are controlled by the proposed algorithm. The analytical thermal model is used to formulate thermal constraints for the system and the cost function of the optimization problem involves pump and chiller plant power consumptions with corresponding weight factors that can be tuned depending on a system's priority. Various heuristics can then be applied to solve this problem dynamically.

References

[1] Gomaa, M., M.D. Powell, and T.N. Vijaykumar, *Heat-and-run: Leveraging SMT and CMP to manage power density through the operating system*. in *ACM SIGARCH Computer Architecture News*. 2004. ACM.

[2] Kursun, E., et al., *Investigating the effects of task scheduling on thermal behavior*. in *Third Workshop on Temperature-Aware Computer Systems (TACS'06)*. 2006.

[3] McDougall, R. and J. Mauro, *Solaris internals: Solaris 10 and OpenSolaris kernel architecture*. 2006. Pearson Education, New York, NY, USA.

[4] Choi, J., et al., *Thermal-aware task scheduling at the system software level*. in *Proceedings of the 2007 International Symposium on Low Power Electronics and Design*. 2007. ACM, Piscataway, NJ, USA.

[5] Seongmoo, H., K. Barr, and K. Asanovic, *Reducing power density through activity migration*. in *Low Power Electronics and Design, 2003. ISLPED '03. Proceedings of the 2003 International Symposium on*. 2003.

[6] Shayesteh, A., et al., *Reducing the latency and area cost of core swapping through shared helper engines*. in *Computer Design: VLSI in Computers and Processors, 2005. ICCD 2005. Proceedings. 2005 IEEE International Conference on*. 2005.

[7] Michaud, P., et al., A study of thread migration in temperature-constrained multicores. ACM Transactions on Architecture and Code Optimization (TACO), 2007. **4**(2): p. 9.

[8] Fabrizio, M., et al., *Thermal balancing policy for streaming computing on multiprocessor architectures*, in *Proceedings of the Conference on Design, Automation and Test in Europe*. 2008. ACM, Munich, Germany.

[9] Zhao, D., H. Homayoun, and A.V. Veidenbaum. *Temperature aware thread migration in 3D architecture with stacked DRAM*. in *Quality Electronic Design (ISQED), 2013 14th International Symposium on*. 2013. IEEE.

[10] Kursun, E. and C.-Y. Cher, Temperature variation characterization and thermal management of multicore architectures. IEEE Micro, 2009(1): pp. 116–126.

[11] Eratne, S., P. Nair, and E. John, A thermal-aware scheduling algorithm for core migration in multicore processors. Journal of Low Power Electronics, 2015. **11**(2): pp. 103–111.

[12] Xuan, S. and Y. Yang, *Temperature-aware thread assignment of many-core processor*. in *Quality Electronic Design (ISQED), 2015 16th International Symposium on*. 2015. IEEE.

[13] Yeo, I., C.C. Liu, and E.J. Kim. *Predictive dynamic thermal management for multicore systems*. in *Proceedings of the 45th Annual Design Automation Conference*. 2008. ACM.

[14] Coskun, A.K., T.S. Rosing, and K.C. Gross, Utilizing predictors for efficient thermal management in multiprocessor SoCs. Computer-Aided Design of Integrated Circuits and Systems, IEEE Transactions on, 2009. **28**(10): pp. 1503–1516.

[15] Xia, L., et al., Implementing a thermal-aware scheduler in linux kernel on a multi-core processor. The Computer Journal, 2010. **53**(7): pp. 895–903.

[16] Li, D., et al., *System-level, thermal-aware, fully-loaded process scheduling.* in *Parallel and Distributed Processing, 2008. IPDPS 2008. IEEE International Symposium on.* 2008. IEEE.

[17] Hanumaiah, V., et al., *Throughput optimal task allocation under thermal constraints for multi-core processors.* in *Design Automation Conference, 2009. DAC '09. 46th ACM/IEEE.* 2009.

[18] Munkres, J., Algorithms for the assignment and transportation problems. Journal of the Society for Industrial and Applied Mathematics, 1957. **5**(1): pp. 32–38.

[19] Ayse Kivilcim, C., et al., *Temperature-aware MPSoC scheduling for reducing hot spots and gradients*, in *Proceedings of the 2008 Asia and South Pacific Design Automation Conference.* 2008. IEEE Computer Society Press, Seoul, Korea.

[20] Coskun, A.K., T.S. Rosing, and K. Whisnant, *Temperature aware task scheduling in MPSoCs.* in *Proceedings of the Conference on Design, Automation and Test in Europe.* 2007. EDA Consortium.

[21] Lu, S.J., R. Tessier, and W. Burleson. *Reinforcement learning for thermal-aware many-core task allocation.* in *Proceedings of the 25th Edition on Great Lakes Symposium on VLSI.* 2015. ACM.

[22] Zhou, X., et al., *Thermal-aware task scheduling for 3d multicore processors.* Parallel and Distributed Systems, IEEE Transactions on, 2010. **21**(1): pp. 60–71.

[23] Chien-Hui, L., H.P.W. Charles, and C. Krishnendu, *An online thermal-constrained task scheduler for 3D multi-core processors*, in *Proceedings of the 2015 Design, Automation & Test in Europe Conference & Exhibition.* 2015. EDA Consortium, Grenoble, France.

[24] Coskun, A.K., et al. *Dynamic thermal management in 3D multicore architectures.* in *Design, Automation & Test in Europe Conference & Exhibition, 2009. DATE'09.* 2009. IEEE.

[25] Yuan, X. and H. Wei-Lun, Temperature-aware task allocation and scheduling for embedded Multiprocessor Systems-on-Chip (MPSoC) design. Journal of VLSI Signal Processing-Systems, 2006. **45**(3): pp. 177–189.

[26] Wu, G., et al., *Task allocation and migration algorithm for temperature-constrained real-time multi-core systems.* in *Proceedings of the 2010 IEEE/ACM Int'l Conference on Green Computing and Communications & Int'l Conference on Cyber, Physical and Social Computing.* 2010. IEEE Computer Society.

[27] Goossens, J. and N. Fisher, The chilling effect of parallelism: Analysis and allocation of parallel real-time jobs for peak system-temperature minimization. Rtsops 2012, **2012**: p. 7.

[28] Baati, K. and M. Auguin, *Temperature-aware DVFS-DPM for real-time applications under variable ambient temperature*. in *Industrial Embedded Systems (SIES), 2013 8th IEEE International Symposium on*. 2013.

[29] Schor, L., et al., *Thermal-Aware Task Assignment for Real-Time Applications on Multi-Core Systems*, in *Formal Methods for Components and Objects*, B. Beckert, et al., Editors. 2013, Springer, Berlin/Heidelberg. pp. 294–313.

[30] Thiele, L., S. Chakraborty, and M. Naedele, *Real-time calculus for scheduling hard real-time systems*. in *Circuits and Systems, 2000. Proceedings. ISCAS 2000 Geneva. The 2000 IEEE International Symposium on*. 2000. IEEE.

[31] Jayaseelan, R. and T. Mitra, *Temperature aware scheduling for embedded processors*. in *VLSI Design, 2009 22nd International Conference on*. 2009. IEEE.

[32] Ghahfarokhi, F.S. and A. Ejlali, *Schedule swapping: A technique for temperature management of distributed embedded systems*. in *Embedded and Ubiquitous Computing (EUC), 2010 IEEE/IFIP 8th International Conference on*. 2010. IEEE.

[33] Uhlig, R., et al., Intel virtualization technology. Computer, 2005. **38**(5): pp. 48–56.

[34] Fortes, J., R. Figueiredo, and P.A. Dinda, Guest editors' introduction: Resource virtualization renaissance. Computer, 2005. **38**(5): pp. 28–31.

[35] Rodero, I., et al., Energy-efficient thermal-aware autonomic management of virtualized HPC cloud infrastructure. Journal of Grid Computing, 2012. **10**(3): pp. 447–473.

[36] Nathuji, R. and K. Schwan, *VirtualPower: Coordinated power management in virtualized enterprise systems*. in *ACM SIGOPS Operating Systems Review*. 2007. ACM.

[37] Khan, O. and S. Kundu, *Predictive thermal management for chip multiprocessors using co-designed virtual machines*, in *High Performance Embedded Architectures and Compilers*. 2009, Springer. pp. 293–307.

[38] Klee, C., *Transparent, thermal balancing of virtual machines in multicore systems*. 2007. Karlsruhe Institute of Technology, Karlsruhe, Germany.

[39] Stoess, J., et al., *Transparent, power-aware migration in virtualized systems*. Proceedings of GI/ITG Fachgruppentreffen Betriebssysteme, 2007: pp. 1–6.

[40] Hung, W., et al., *Thermal-aware IP virtualization and placement for networks-on-chip architecture*. in *Computer Design: VLSI in Computers and Processors, 2004. ICCD 2004. Proceedings. IEEE International Conference on*. 2004. IEEE.

[41] Ayala, J.L., C. Méndez, and M. López-Vallejo, *Analysis of the thermal impact of source-code transformations in embedded-processors*. in *Electronics, Circuits and Systems, 2006. ICECS'06. 13th IEEE International Conference on*. 2006. IEEE.

[42] Mukherjee, R., S.O. Memik, and G. Memik, *Temperature-aware resource allocation and binding in high-level synthesis*. in *Proceedings of the 42nd Annual Design Automation Conference*. 2005. ACM.

[43] Ni, M. and S.O. Memik, *Thermal-induced leakage power optimization by redundant resource allocation*. in *Proceedings of the 2006 IEEE/ACM International Conference on Computer-Aided Design*. 2006. ACM.

[44] Mutyam, M., et al., *Compiler-directed thermal management for VLIW functional units*. in *ACM SIGPLAN Notices*. 2006. ACM.

[45] Cameron, K.W., H.K. Pyla, and S. Varadarajan, *Tempest: A portable tool to identify hot spots in parallel code*. in *Parallel Processing, 2007. ICPP 2007. International Conference on*. 2007.

[46] Kremer, U., *Compilation Techniques for Power, Energy, and Thermal Management*, in *Designing Embedded Processors*. 2007, Springer. pp. 287–303.

[47] Lin, T.-Y., C.-Y. Lee, and R.-G. Chang, *Thermal-aware register allocation with loop-unrolling*. in *Computer Design and Applications (ICCDA), 2010 International Conference on*. 2010. IEEE.

[48] Rodero, I., et al., *Investigating the potential of application-centric aggressive power management for HPC workloads*. in *High Performance Computing (HiPC), 2010 International Conference on*. 2010. IEEE.

[49] Tang, Q., et al., *Sensor-based fast thermal evaluation model for energy efficient high-performance datacenters*. in *Intelligent Sensing and Information Processing, 2006. ICISIP 2006. Fourth International Conference on*. 2006. IEEE.

[50] Moore, J.D., et al., *Making scheduling "cool": Temperature-aware workload placement in data centers*. in *USENIX Annual Technical Conference, General Track*. 2005.

[51] Moore, J., J.S. Chase, and P. Ranganathan, *Weatherman: Automated, online, and predictive thermal mapping and management for data centers*. in *3rd International Conference on Autonomic Computing, Proceedings*, 2005: pp. 155–164.

[52] Nemati, K., B.T. Murray, and B. Sammakia, *Experimental characterization and modeling of a water-cooled server cabinet*. in *Thermal and Thermomechanical Phenomena in Electronic Systems (ITherm), 2014 IEEE Intersociety Conference on*. 2014. IEEE.

[53] Moore, J., et al. *Going beyond CPUs: The potential of temperature-aware solutions for the data center*. in *Proc. 2004 1st Workshop Temperature-Aware Computer Systems (TACS-1) Held in Conjunction with ISCA-31*. 2004.

[54] Bash, C. and G. Forman. *Cool job allocation: Measuring the power savings of placing jobs at cooling-efficient locations in the data center*. in *USENIX Annual Technical Conference*. 2007.

[55] Tang, Q., S.K.S. Gupta, and G. Varsamopoulos, Energy-efficient thermal-aware task scheduling for homogeneous high-performance computing data centers: A cyber-physical approach. Parallel and Distributed Systems, IEEE Transactions on, 2008. **19**(11): pp. 1458–1472.

[56] Li, L., et al., *Thermocast: A cyber-physical forecasting model for datacenters*. in *Proceedings of the 17th ACM SIGKDD International Conference on Knowledge Discovery and Data Mining*. 2011. ACM.

[57] Mukherjee, T., et al., *Software architecture for dynamic thermal management in datacenters*. in *Communication Systems Software and Middleware, 2007. COMSWARE 2007. 2nd International Conference on*. 2007. IEEE.

[58] Rodero, I., et al., *Towards energy-efficient reactive thermal management in instrumented datacenters*. in *Grid Computing (GRID), 2010 11th IEEE/ACM International Conference on*. 2010. IEEE.

[59] Abbasi, Z., G. Varsamopoulos, and S.K.S. Gupta, *Thermal aware server provisioning and workload distribution for internet data centers*. in *Proceedings of the 19th ACM International Symposium on High Performance Distributed Computing*. 2010. ACM.

[60] Rao, T.V.N., et al., Sustainable environment friendly green computing for futuristic technologies. Journal of Information Sciences and Computing Technologies, 2015. **2**(1): pp. 128–132.

[61] Shoukourian, H., et al., Towards a unified energy efficiency evaluation toolset: An approach and its implementation at Leibniz supercomputing centre (lrz). Information and Communication Technologies, 2013: p. 276.

[62] Conficoni, C., et al., *Energy-aware cooling for hot-water cooled supercomputers*. in *Proceedings of the 2015 Design, Automation & Test in Europe Conference & Exhibition*. 2015. EDA Consortium.

Chapter 6
Emerging directions in thermal-aware systems

The impact of thermal effects have first been noted due to their direct relationship with the power consumption, performance, reliability, and cooling needs of the integrated circuits deployed in high-performance systems. With a growing awareness of this phenomenon, various other aspects are now emerging as important design considerations. A striking example is how users physically perceive heat dissipated from electronic devices, particularly, users of mobile and handheld devices. Another dimension to the issue has been emerging as the amount of heat generated by some of these systems becomes so substantial that we now speak of harvesting this energy to convert back to electric energy. Thermal harvesting is a vast domain with a plethora of applications in systems at various macro and micro scales. It is beyond the scope of this book to provide a thorough discussion of this topic. In this chapter a brief introduction to recent development specifically focusing on on-chip thermal harvesting will be presented. Another avenue of innovation is the pursuit of novel materials and design methods for building on-chip temperature monitoring devices. Finally, the presence of heat and the mechanisms that generate heat within integrated circuits can be viewed from an entirely different perspective to infer valuable knowledge on the structure and operation of an integrated circuit. Securing integrated circuits and understanding potential threats that may exploit thermal properties is one such interesting domain worth mentioning in the conclusion of this book.

6.1 Considering user comfort in thermal design

Skin temperature of mobile devices intimately affects the user experience. Power management schemes built into handheld devices such as smartphones can lead to quickly crossing a user's threshold of tolerable skin temperature. Furthermore, there is a significant variation among users in terms of their sensitivity. Users are often in touch with the back cover and screen for prolonged durations. Thus, heat dissipation intimately affects user experience. Heat pain is experienced by most humans when they touch an object hotter than 45°C [1, 2], which is not uncommon in mobile devices. There have been cases reported regarding the elevated outer temperatures of mobile devices while playing graphics intensive games and it is reported that the hottest spot reaches 47°C [3]. In fact, excessive temperatures are common to many smartphones [4, 5].

The need for more sophisticated thermal management for smartphones, including special consideration to the user experience, has been recognized in industry [6]. Mitigation of the thermal challenges has been first attempted by improving the thermo-mechanical properties of the devices. The effects of different materials and form factors used for the casing of a smartphone has been investigated, for instance Reference 7. The thermal response of various casing geometries under different conditions (e.g., in free air, held by a person, placed inside a garment's pocket, etc.) has been evaluated using power consumption ranges representative of smartphone workloads.

Other studies aimed at generating equivalent thermal models of these devices and their associated thermal simulations and modeling environments. One study proposed a thermal analyzer, which generates steady-state temperature maps of entire smartphone including the cover [8]. Another study demonstrates the thermal coupling between the application processor and the battery [9].

More recent developments occurred towards better system-level management of smartphones to provide better user experience that does not require any hardware design change. A study demonstrated the variability among smartphone users by means of temperature perception [10]. In this study, experiments were conducted with a commercially available smartphone running Android OS. This smartphone has two built-in thermal sensors to report the CPU and battery temperatures. This set was extended with two sensors on the back cover (upper section and midsection) and one on the screen and system-level information, such as CPU temperature, battery temperature, CPU utilization, and CPU frequency were logged periodically with an application.

A user study with 10 participants (5 males, 5 females), who are graduate or undergraduate students at a university was then performed. Subjects were asked to hold the phone in their hands and to report the instant when the discomfort due to skin temperature was unacceptable to them. The most tolerant subject ended the test in 7 min. This is consistent with a realistic environment, where an average user could easily be expected to touch the smartphone for several minutes. A computationally heavy benchmark application was used during the user study. Even though the CPU temperature does not exceed the threshold for the built-in power management system to intervene, it was observed that this very same thermal state causes the skin temperature to exceed comfortable limits for all the participants. Another important result of this study is the sensitivity variation across users. Figure 6.1 shows that the minimum skin temperature exceeding the comfort limit for a user is 34.0°C, while the maximum is 42.8°C.

A system-level management scheme for user-aware skin temperature control has then been proposed targeting this demonstrated need [10]. The first component is a system-level skin temperature prediction model. The prediction system is based on empirical data gathered from thermal measurements as well as performance indicators accessible through the Android OS. Skin and screen temperature measurements were collected from an instrumented smartphone with attached external thermistors. These temperature readings were used as the expected value for the skin temperature and screen temperature. Simultaneously, the CPU temperature,

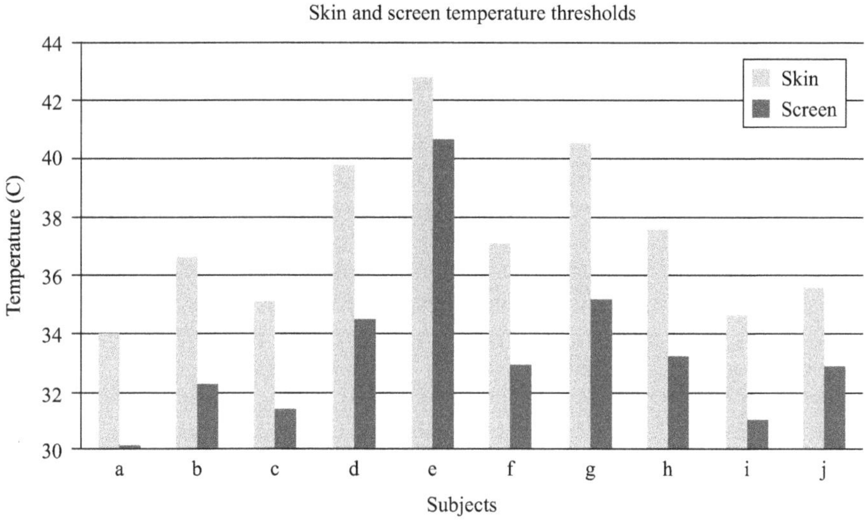

Figure 6.1 A user study indicating the extent of variation among individual levels of comfort with respect to skin and screen temperatures for smartphone users [10]

CPU utilization and clock frequency, as well as battery temperature were collected from the OS in a log file. Using the log data as the training data, various machine learning methods have been used to construct run-time skin temperature models, including linear regression, multilayer perceptron, M5P, and REPtree. Ten-fold cross validation performed on all models reveals that linear regression and multilayer perceptron are relatively poor in accuracy when compared to other methods. Accuracy is measured as the percentage error between the expected skin temperature (as derived from instrumented external sensors) and the predicted skin temperature. M5P and REPtree make nearly 100% accurate estimations. REPtree gives the best results with average of 0.95% error for skin temperature estimations. M5P algorithm has 0.96% error. However, when temperature differences that are less than 1°C (as humans are less sensitive in that range) are ignored, the M5P algorithm ultimately gives better results and it has been implemented within the Android OS as part of the dynamic power management module. This study, then evaluates a use case for such a predictor by introducing it into the feedback loop of the DVFS mechanism of the phone's power governor.

The model is used by the DVFS system to keep skin temperature under a user's desired threshold, thereby minimize discomfort. This particular study experimented with both using a default threshold representative of an average user as well as configuring it for each individual's threshold with an application interface. The management scheme performs skin temperature prediction every 3 s and intervenes to enforce a DVFS decision on the system only if skin temperature needs to be controlled. Otherwise, the baseline DVFS of the smartphone's OS performs its function for power optimization only. The baseline DVFS is the default Android on-demand governor and it scales the frequency of the processor according to CPU utilization. When utilization is at the maximum, the frequency is also set at the maximum level. The reduction in frequency can be steep if the utilization is very low or it could be in steps if the utilization is below a threshold (around 80%), but above a minimum (around 20%). For a typical smartphone used in this study, there are twelve frequency levels between a few hundred MHz and slightly under 2 GHz.

The user comfort aware skin temperature control mechanism has a threshold for activation which is set to 2°C below the skin temperature limit of the user. If the difference between the predicted skin temperature and the temperature limit is between 1°C and 2°C, the maximum allowed CPU frequency is decreased by one level (i.e., from the highest frequency to the one below). If the difference between the prediction and the temperature limit is between 0.5°C and 1°C, then, the maximum allowed CPU frequency is decreased by two levels. Finally, if the prediction is closer than 0.5°C to the limit or it is exceeding the limit, then, the maximum CPU frequency is set to the minimum frequency level. In user studies performed with this new system, the skin temperature aware scheme has been favored by users by a large margin while the same users finding the performance of the smartphone acceptable at the same time.

6.2 Thermal harvesting from within integrated circuits

Conversion of heat into electric energy is a well-known concept with macro-scale applications. On the other hand, on-chip thermal harvesting at the micro and nano scales is an emerging domain. Thermal energy resulting from hotspots within ICs can present opportunities for harvesting. Especially 3D ICs exhibit long lived temperature gradients across the lateral surfaces that are almost guaranteed to always exist. Some thermal harvesting schemes have been proposed in an attempt to take advantage of the thermal output in today's high-performance chips [11–13]. Bulk thermoelectric materials such as those discussed for temperature sensor applications in Chapter 2 can be used for this purpose as well. However, existing approaches aim at thermal harvesting at the macro scale, either as discrete components attached outside the case or as off-chip components underneath the chip package. Thin-film and nanostructured thermoelectric devices prove to be more suitable for on-chip integration than bulk discrete devices.

One of these approaches proposed two thermal harvesting schemes, inserting a thermoelectric module between the CPU and heat spreader or placing the thermoelectric module on an extended heat spreader [12]. Another one proposed a thermal harvesting system by integrating a 180 μm thick N-P superlattice structured thermoelectric module between the heat sink and the die [11]. In this configuration, the harvester can only exploit the gradient between the chip and the heat sink, which inherently is expected to be small, since the purpose of the heat sink is to equalize the thermal state at this interface. Also, addition of a component into the cooling path between the die and heat sink will have a negative impact on the overall cooling efficiency. An off-chip harvester inserted between a die and the package has been investigated, mainly focusing on the design of the power regulator circuitry [14]. However, this study does not discuss any particular way of using the harvested energy and what the capabilities of this system would be end-to-end. A multi-input platform for harvesting was investigated, yet, this design also does not provide any specifics as to how the harvester system interacts with actual load components [15]. A demonstration of harvesting was presented using the output power of a bulk thermoelectric device by attaching it to a CPU package [13]. N-P thermoelectric modules are also offered as commercially available products. However, these devices are not on-chip integrated, instead they are placed between the die and the heat sink.

A more recent study aims to establish an end-to-end analysis of a nano scale on-chip thermal harvester [16]. A typical thermal energy harvesting system consists of three main stages: thermoelectric converter, voltage converter, and a charge buffer. One realization of this harvesting system using bi-metallic thermocouples is shown in Figure 6.2 [16]. A bi-metallic thin-film thermocouple array converts thermal energy into electrical energy. The second stage consists of a power converter to collect and boost the low voltage converter output. The third stage is the energy buffer, which is an on-chip capacitor used to store energy and power a load. There is also a control system that produces clock signal Φ/Φ' for the power converter circuitry. The control system also charges and discharges the energy buffer based on the usage of the load.

A bi-metallic thin-film thermocouple consists of two metal layers, connected electrically at one junction and open at the other junction. Any temperature gradient experienced by this device will create a V_{HC} (hot-to-cold voltage) proportional to the resulting Seebeck coefficient of the two metal layers. When the thermocouples are connected electrically in series, the final output voltage is a summation of individual thermocouple output voltages. Figure 6.3 shows an illustration of a thermal harvester placement on a die. The die has a thermal gradient, ΔT, from one end to the opposite end. The thermocouple array is placed in parallel to the direction of the thermal gradient so that the maximum temperature gradient can be created between the two junctions of each thermocouple. Total V_{HC} is the summation of the output voltages of the individual thermocouples connected in series. Each thermocouple consists of two metals, shown in Figure 6.3 as Metal 1 and Metal 2, respectively. The connection between neighboring devices is made by

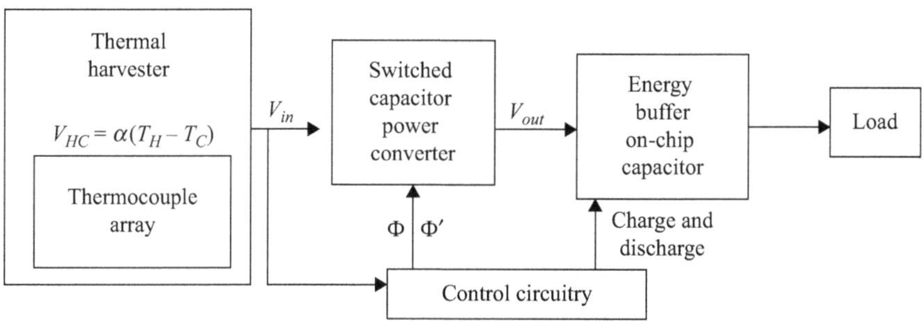

Figure 6.2 Overall architecture of an on-chip thermal harvesting system

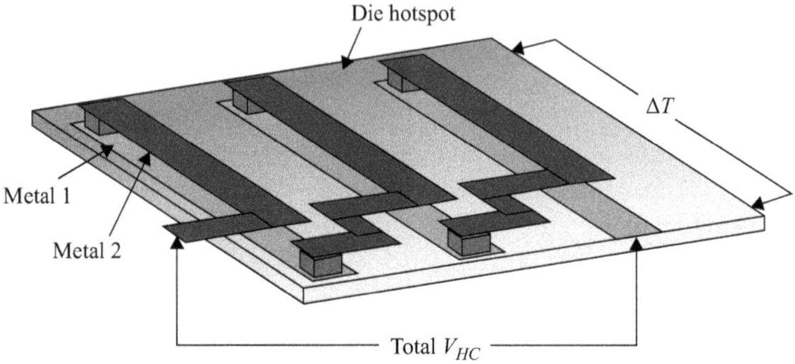

Figure 6.3 Thermal harvester: the output voltage of each thin-film thermocouple device is summed to total V_{HC}

extending the wire from one device's Metal 1 layer through a via to the next device's Metal 2 layer.

Individual thin-film thermocouple devices produce low output voltages. However, the actual thermal harvester will be a large array of thermocouples connected in series and will produce output in the range of hundreds of mV. This voltage output can be first boosted through a converter circuit and then stored into a charge buffer. A switched capacitor converter design [14], which allows optimization of the number of stages according to the specific input and output voltage levels can be used for this purpose. The optimization relation is shown below in (6.1), where N_{op} is the optimal number of stages for the power converter for a specific input and output voltage. For example, if V_{in} is 0.3 V (the output of the thermocouple array) and V_{out} is set to 1.0 V, this will result in a 5-stage converter.

$$N_{op} = 2 * \left(\frac{V_{out}}{V_{in}} - 1\right) \tag{6.1}$$

The output of this power converter is connected to an energy buffer, which can be built using an on-chip capacitor. The energy buffer is the last stage in a thermal harvesting system. This stage is responsible for collecting and saving all the energy from the source. At the appropriate period, the energy buffer will release the stored charge into the load application. For a complete on-chip integration of a thermal harvesting system, it is necessary that the energy buffer is of reasonable size, close to the form factor of the die. At the same time the energy buffer needs to be able to store enough charge to power a useful block of circuitry. They can be built similar to on-chip decoupling capacitors. Typical on-chip decoupling capacitor values range from 1 pF to 162 pF [17]. Hence, it is plausible to embed a charge buffer of similar dimensions for a harvesting system.

6.3 New materials and designs for on-chip temperature sensors

New materials are under investigation as an alternative to CMOS-based integrated circuit components. Among these, spintronic devices are attracting significant attention. There are also investigations of building temperature sensors using this emerging technology. An early proposal investigated the impact of temperature on spintronic memristor devices and identified two distinct regions of operation: temperature-sensitive and temperature-insensitive regions of operation [18]. A property of this device called the domain-wall mobility depends on both the degree of spin torque excitation and the thermal fluctuation experienced by the device. After biasing a spintronic memristor device at a constant voltage pulse, a change in its resistance can be observed. Furthermore, at a higher temperature, the variation in resistance before and after the pulse is larger. This relationship can be exploited to use the device as a temperature sensor.

Another proposal exploits the relationship between a memristor's resistance and temperature [19]. This design combines CMOS-based current biasing circuitry with a small footprint memristor device to measure temperature-induced resistance fluctuations.

Another recent proposal involves a temperature sensing resistive random access memory to perform temperature monitoring [20]. This is an interesting proposal to use a single bit memory both as a memory device and a temperature sensor. The architecture is envisioned as a 3D stacked memory and logic integration. The resistive random access memory layer performs the thermal monitoring of the hotspots that occur within the underlying logic layer.

6.4 Hardware security

Another interesting aspect related to heat generated within integrated circuits is what that heat output may indicate as far as the functionality of that circuit is concerned. One application of this concept is on hardware Trojan detection via temperature tracking. One recent proposal leverages on-chip temperature sensors for run-time Trojan detection [21, 22]. Deviations in the power and thermal profile of an integrated circuit can be exploited as indicators of unplanned/foreign introduction of components, such as hardware Trojans. Since these malicious structures are usually designed to be triggered for specific and rare events post-deployment, it is quite challenging to perform the appropriate testing procedure to excite and uncover them at testing. Therefore, continuous monitoring at run-time is extremely valuable, which temperature sensors can provide. The proposed method, in fact both operates at the time of testing to detect early attacks and isolate compromised chips and it also performs continuous monitoring. Using thermal profiles of a chip derived during testing a model for expected correlation between different individual temperature sensors in a given chip is first derived. Next, at run-time, this correlation is repeatedly validated using a Kalman filter.

Another study investigated temperature tracking-based power consumption attacks, which are a type of the so-called side-channel attacks aimed at extracting secure information from a protected hardware structure [23]. Specifically, in this study, the main goal is to secure the task execution ordering of a real-time embedded system. The main observation is that different orders of task executions can have distinctly different thermal profiles, enabling the attacker to infer the task execution sequence using thermal tracking-based power consumption attacks. If this information is extracted from the system, the system can become highly vulnerable to future attacks specifically configured to disable or disrupt this known task sequence. As a possible defense mechanism against an adversary executing a malicious task sequence, a task scheduling mechanism, which leaks the smallest possible amount of side-channel information is proposed. This algorithm models the behavior of a powerful attacker and then executes a counter task schedule that trades-off information leakage against delays introduced into the schedule and resulting delays in meeting task deadlines.

Another method for detecting hardware Trojans proposed to perform 2D principle component analysis on run-time thermal and power maps is the presence of process variation [24]. The diagnosis framework extracts detailed thermal maps of chips using infrared imaging and then re-constructs fine grain power maps using thermal to power inversion procedures. It is plausible to greatly reduce the cost of this test in the future with the availability of orders of magnitude higher amounts of on-chip temperature sensors creating an extremely fine-grain monitoring network.

References

[1] Arens, E.A. and H. Zhang, *The skin's role in human thermoregulation and comfort*. 2006. Center for the Built Environment, University of California Berkeley, Berkeley, CA, USA.

[2] Wasner, G.L. and J.A. Brock, Determinants of thermal pain thresholds in normal subjects. Clinical Neurophysiology, 2008. **119**(10): pp. 2389–2395.

[3] Kaplan, J.A., *New Apple iPad Hits 116 degrees, Consumer Reports Says*. FoxNews.com, 2012.

[4] Herman, J., *Why is My Phone So Hot*. Popular Mechanics. Retrieved from: http://www.popularmechanics.com/technology/how-to/tips/why-does-my-phone-get-so-hot, 2011.

[5] *Sony Xperia Z2 Smartphone Gets Too 'Hot' to Use*. Available from: http://www.maxconsole.com/news/ANDROID, 2014.

[6] Sekar, K. *Power and thermal challenges in mobile devices*. in *Proceedings of the 19th Annual International Conference on Mobile Computing & Networking*. 2013. ACM.

[7] Lee, J., D.W. Gerlach, and Y.K. Joshi. *Parametric thermal modeling of heat transfer in handheld electronic devices*. in *Thermal and Thermomechanical Phenomena in Electronic Systems, 2008. ITHERM 2008. 11th Intersociety Conference on*. 2008. IEEE.

[8] Xie, Q., M.J. Dousti, and M. Pedram. *Therminator: A thermal simulator for smartphones producing accurate chip and skin temperature maps*. in *Proceedings of the 2014 International Symposium on Low Power Electronics and Design*. 2014. ACM.

[9] Xie, Q., et al., *Dynamic thermal management in mobile devices considering the thermal coupling between battery and application processor*. in *Computer-Aided Design (ICCAD), 2013 IEEE/ACM International Conference on*. 2013. IEEE.

[10] Egilmez, B., et al., *User-specific skin temperature-aware DVFS for smartphones*. in *Proceedings of the 2015 Design, Automation & Test in Europe Conference & Exhibition*. 2015. EDA Consortium.

[11] Choday, S.H., et al., *On-chip energy harvesting using thin-film thermoelectric materials.* in *Semiconductor Thermal Measurement and Management Symposium (SEMI-THERM), 2013 29th Annual IEEE.* 2013. IEEE.

[12] Zhou, Y., S. Paul, and S. Bhunia. *Harvesting wasted heat in a microprocessor using thermoelectric generators: Modeling, analysis and measurement*. in *Proceedings of the Conference on Design, Automation and Test in Europe.* 2008. ACM.

[13] Fahad, H., et al., Thermoelectricity from wasted heat of integrated circuits. Applied Nanoscience, 2013. **3**(3): pp. 175–178.

[14] Lu, C., et al., *Stage number optimization for switched capacitor power converters in micro-scale energy harvesting. in Design, Automation & Test in Europe Conference & Exhibition (DATE)*. 2011. IEEE.

[15] Bandyopadhyay, S. and A.P. Chandrakasan, *Platform architecture for solar, thermal, and vibration energy combining with MPPT and single inductor.* Solid-State Circuits, IEEE Journal of, 2012. **47**(9): pp. 2199–2215.

[16] Li, D., S. Ogrenci-Memik, and L. Henschen. *On-chip integration of thermoelectric energy harvesting in 3D ICs*. in *Circuits and Systems (ISCAS), 2015 IEEE International Symposium on*. 2015. IEEE.

[17] Popovich, M., A.V. Mezhiba, and E.G. Friedman, Efficient placement of distributed on-chip decoupling capacitors. Power Distribution Networks with On-Chip Decoupling Capacitors, 2008: pp. 435–458.

[18] Xiaobin, W., et al., Spintronic memristor temperature sensor. *Electron Device Letters*, IEEE, 2010. **31**(1): pp. 20–22.

[19] Bi, X., et al., *Spintronic memristor based temperature sensor design with cmos current reference.* in *Proceedings of the Conference on Design, Automation and Test in Europe.* 2012. EDA Consortium.

[20] Merkel, C.E. and D. Kudithipudi, Temperature sensing RRAM architecture for 3-D ICs. *Very Large Scale Integration (VLSI) Systems, IEEE Transactions on*, 2014. **22**(4): pp. 878–887.

[21] Domenic, F., B. Chongxi, and S. Ankur, *Temperature tracking: An innovative run-time approach for hardware Trojan detection,* in *Proceedings of the International Conference on Computer-Aided Design.* 2013, IEEE Press: San Jose, California.

[22] Forte, D.J., *Design, Fabrication, and Run-time Strategies for Hardware-Assisted Security, PhD dissertation*. 2013. University of Maryland, College Park, MD, USA.

[23] Chongxi, B. and S. Ankur, *A secure algorithm for task scheduling against side-channel sttacks*. in *Proceedings of the 4th International Workshop on Trustworthy Embedded Devices*. 2014. ACM, Scottsdale, Arizona, USA.

[24] Nowroz, A.N., et al., *Novel techniques for high-sensitivity hardware Trojan detection using thermal and power maps*. in *Computer-Aided Design of Integrated Circuits and Systems, IEEE Transactions on*, 2014. **33**(12): pp. 1792–1805.

Appendix A

Relevant units and metrics

Heat	Joule
Power consumption	W
Power density	Wcm^{-2}
Temperature	K (°C)
Temperature sensor sensitivity	$V°C^{-1}$ for voltage output ($A°C^{-1}$ for current output)
Temperature sensor linearity	defined by a range of temperatures or offset (°C)
Temperature sensor resolution	°C (number of bits of the ADC)
Thermal capacitance	JK^{-1}
Thermal conductivity	$Wm^{-1}K^{-1}$
Thermal diffusivity	m^2s^{-1}
Thermal resistance	KW^{-1}
Thermal time constant	rate of rise from ambient temperature in response to a step input of power dissipation
Thermistor temperature coefficient	$ppm°C^{-1}$
Seebeck coefficient	$V\ K^{-1}$ ($V°C^{-1}$)
Volumetric heat capacity	$Jm^{-3}K^{-1}$

Index

www.ingramcontent.com/pod-product-compliance
Ingram Content Group UK Ltd.
Pitfield, Milton Keynes, MK11 3LW, UK
UKHW020803290926
15470UKWH00003B/924